·AN ILLUSTRATED GUIDE·

JAMES SAUNT

Sinclair International Limited, Norwich, England
1990

FOR MY LATE MOTHER KATHLEEN, AND TO MY FATHER ROLAND;
JAN, CATHERINE, DEBORAH AND MICHAEL.

Acknowledgements

Friends and associates in many countries were generous in providing fruit samples for photography or information on local varieties. The invaluable assistance given by the following is gratefully acknowledged:

Argentina: A. H. Gasparri, Buenos Aires, E. C. Taylor, Citricola Ayui, Concordia; **Australia:** E. Benham, Mundubbera, Queensland, D. J. P. McIntyre, Gayndah, Queensland, J. W. Turpin, Sydney, New South Wales; **Brazil:** C. Moreira, Piracicaba, Sao Paulo, G. Muller, Campinas, Sao Paulo; **China:** Cheong-Yin Wong, Guangzhou, Lui Xiao Zhong, Chongquing, Sichuan; **Cyprus:** A. J. Christodoulou, Lanitis Farms, Limassol, G. Tsimon, Phassouri Plantations, Limassol; **England:** L. Calcutt, SACCE, London; **France:** P. Cardon, SACCE, Paris; **Israel:** M. Davidson and M. Rahat, CMBI, Tel Aviv; **Italy:** V. Lo Giudice, ISA, Acireale, Sicily; **Japan:** B. H. Houlden, Gold Reef, Tokyo, J. Ishida, Matsuyama, Ehime, Y. T. Yamaki, Ninomiya-Machi; **Netherlands:** H. van Dongen, SACCE, Barendrecht; **New Zealand:** A. R. Harty, Kerikeri; **Saudi Arabia:** W. T. Walker, Sharbatley Corp, Jeddah; **South Africa:** C. J. Alexander, SACCE, Port Elizabeth, S. Burdette, OCC, Nelspruit, A. T. C. Lee, Port Elizabeth, L. von Broembsen, Pretoria; **Spain:** R. Bono Ubeda, IVIA, Moncada, Valencia, A. Garcia Lidon, Murcia, M. Ricart Vila, Albal, Valencia, I. Sanchez Martorell, Pasqual Hermanos, Valencia; **Swaziland:** H. C. Noddeboe, Ngonini, Piggs Peak, K. R. Pyle, Inyoni Yami, Tshanei; **Turkey:** S. Hayirlioğlu, Mersin, M. Kaplankiram, University of Çukurova, Adana; **United States:** G. E. Carman, U.C. Riverside, California, C. J. Hearn, Orlando, Florida, G. Mixon, Bradenton, Florida, D. Swietlik, Weslaco, Texas; **Uruguay:** J. C. Codina, Montevideo, S. Pirez Lostao, Paysandu

I am particularly indebted to the following who gave the benefit of their considerable experience in citricultural matters when reading through the draft manuscript in full or in part, and offering advice and constructive comment. I am most grateful for their help and support: S. Burdette, Nelspruit, South Africa; J. W. Cameron, Riverside, California; W. C. Castle, Lake Alfred, Florida; W. C. Cooper, Winter Park, Florida; F. Gmitter, Lake Alfred, Florida; W. Grierson, Lake Alfred, Florida; E. H. Nauer, Riverside, California; E. Rabe, Nelspruit, South Africa; A. F. G. Smith, Berkhamsted, England; P. Spiegel-Roy, Rehovot, Israel; W. F. Wardowski, Lake Alfred, Florida

I owe Nigel Cattlin (Holt Studios, Hungerford, England) special thanks for photographing the citrus over the three years it took to arrange for them to be collected. Without his efforts this book would be the poorer. I should also like to express my sincere thanks to the following for making available the photographs which appear on the pages indicated: S. Burdette, OCC, Nelspruit, South Africa (p. 85); G. Gamliel, CMBI, Tel Aviv, Israel (pp. 104–5, 109); P. Gooch, Foods From Spain, London (p. 10); E. Hamilton-Harding, Outspan, Berkhamsted, England (pp. 33, 67, 119); M. Johnson, Sunkist Growers Inc, Ontario, California (p. 9); J. van der Walt, Ngonini, Swaziland (p. 97).

Finally, I should like to express my sincere appreciation and gratitude to the Directors of Sinclair International Ltd and especially Peter Briggs, for the backing and encouragement they have provided in the preparation and publication of this book.

Although I have been associated with citrus for more than 30 years several publications, the more important of which are listed on page 126, have proved invaluable in the preparation of this book.

First published in 1990 by Sinclair International Limited,
40 Hellesdon Park Road, Hellesdon Hall Industrial Park, Norwich, NR6 5DR, England

Designer: Janet James Editor: Kate Truman Production: Amanda Newton

Jacket photography by Nigel Cattlin. *Back cover:* Orange trees in the Souss Valley, Morocco.

A CIP catalogue record for this book is available from the British Library.

ISBN 1 872960 00 6

Filmset by Ace Filmsetting Ltd, Frome, Somerset
Colour reproduction by Columbia Offset, Singapore
Printed and bound in Great Britain by BPCC Paulton Books Ltd

CONTENTS

Table 1
World Citrus Production and Utilisation 1986–7
(thousands of tons)

	Total citrus	*Exports (fresh fruit)*	*Processed*
NORTHERN HEMISPHERE	40,773	5,970	10,896
United States	10,832	910	7,180
Mediterranean region	15,794	4,528	2,478
Greece	1,084	225	176
Italy	3,864	327	841
Spain	3,852	2,201	339
Israel	1,453	568	871
Algeria	340	15	
Morocco	984	470	48
Tunisia	295	52	
Cyprus	329	235	62
Egypt	1,506	155	11
Lebanon	315		
Turkey	1,286	152	130
USSR	300		
Japan	2,884		
Cuba	730	453	
Mexico	2,311		
China	2,610	15	
SOUTHERN HEMISPHERE	22,248	813	9,198
Argentina	1,520	135	488
Brazil	15,839	99	8,197
Uruguay	200	68	6
Venezuela	392		
Australia	593	51	344
South Africa	685	435	155
WORLD	63,021	6,783	20,094

Table 2
World Citrus Production by Types 1986–7
(thousands of tons)

	Oranges	*Mandarins*	*Lemons and Limes*	*Grapefruit and Pummelos*
NORTHERN HEMISPHERE	24,965	7,004	4,939	3,864
United States	7,367	363	986	2,325
Mediterranean region	9,878	2,896	2,371	649
Greece	838	68	172	
Italy	2,424	550	882	8
Spain	2,023	1,167	645	18
Israel	870	132	63	388
Algeria	220	105		15
Morocco	650	311	20	4
Tunisia	151	85	25	33
Cyprus	172		55	96
Egypt	1,235	117	152	
Lebanon	250			
Turkey	706	300	250	30
USSR		300		
Japan	341	2,542		
Cuba	410	30	70	200
Mexico	1,480	131	609	
China	2,028	260	123	
SOUTHERN HEMISPHERE	19,557	1,094	1,055	542
Argentina	700	260	400	160
Brazil	14,908	600	295	36
Chile			70	
Uruguay	95	50	48	7
Venezuela	391			
Australia	494		38	30
South Africa	515		60	110
WORLD	44,522	8,099	5,994	4,406

Source: FAO Annual Statistics 1988

INTRODUCTION

Citrus can rightly be regarded as a universal fruit with production in over 100 countries in all six continents. Furthermore, citrus is the most important tree fruit crop in the world, with current world production far exceeding that of all deciduous tree fruits (apples, pears, peaches, plums, etc.). Distribution is in a belt spreading approximately 40° latitude on each side of the Equator and is to be found in tropical and sub-tropical regions where favourable soil and climatic conditions occur. The majority of commercial citrus production, however, is restricted to two narrower belts in the sub-tropics roughly between 20° and 40° N and S of the Equator.

PRESENT-DAY CITRUS PRODUCTION

The area planted to citrus has been estimated at 2 million ha and present-day production of all types at 63 million tons, of which 71 per cent are oranges, 13 per cent mandarins, 9 per cent lemons and limes and 7 per cent grapefruit. Table 1 (see opposite) shows total citrus production for the major producing countries and the quantity exported fresh and processed for each in the 1986–7 season.

Until very recently the United States led world production but this has been overtaken by Brazil. Together these two countries produce about 42 per cent of the world's citrus crop, with 10.8 million tons and 15.8 million tons in the USA and Brazil respectively. Important though fresh fruit sales are in both countries, the majority of their citrus crop is processed: 52 per cent in Brazil and 66 per cent in the USA. Moreover, their combined tonnage of processed citrus accounts for about 75 per cent of the world's total. About 33 per cent of the world's citrus output is processed, mostly into frozen concentrated juice.

Many of the rest of the world's important citrus-producing countries such as Japan, Mexico, Egypt and Argentina have important local markets for fresh citrus fruit, while others such as Spain, Morocco, Israel, Cuba and South Africa depend heavily on exports of fresh fruit as an outlet for much of their production.

Almost all the world's trade in fresh citrus is in or to the Northern Hemisphere, particularly Western Europe. Much of it is from nearby countries in the Mediterranean region, such as Spain, Israel, Morocco, Italy, Cyprus and Greece. The USA has a significant trade to neighbouring Canada and the Far East, particularly to Japan and Hong Kong. The most important exporting country in the Southern Hemisphere is South Africa. Table 2 (see opposite) presents data for the major citrus types (e.g. oranges, lemons, etc.) for the leading producing countries.

Oranges are by far the most extensively produced citrus fruit, accounting for 71 per cent of total production at 44.5 million tons. Production of oranges during the past decade increased by 30 per cent and is forecast to continue to rise at 1.5 per cent per annum to the year 2000. Brazil's 14.9 million ton orange crop represents almost one-third of global production and, when combined with the 7.4 million tons produced annually by the United States, it can be seen that these two leading producers account for slightly more than the rest of the world's combined output. This situation results from the heavy commitment of these two countries to producing a majority of their oranges exclusively for processing into juice.

In the Mediterranean region, Italy and Spain have substantial orange production with 2.4 million and 2.8 million tons respectively. The varieties grown by both industries are primarily for fresh fruit consumption. However, whereas Spain has developed navel orange and other popular varieties, Italy's crop is made up largely of inferior, older varieties for which there is little present-day demand. Consequently, much of Italy's orange production has to be processed or sold into EC intervention and destroyed; less than 1 per cent is exported fresh. Spain, on the other hand, has a thriving export trade and less than 10 per cent is processed.

China's orange crop has traditionally been small since mandarins have been preferred but with a policy change in the late 1960s orange production has increased from an estimated 0.35 million tons in 1975 to over 2 million in 1986. Virtually all China's orange crop is consumed locally. The majority of orange production in Mexico and Egypt is likewise sold fresh on the local market.

There have been significant changes in the world production of mandarins in the recent past. Total production since the mid-1970s increased from 6 to 7 million tons in 1986–7. However, the pattern of mandarin production in Japan and Mediterranean regions – the world's main production areas – is quite different. Japan's overproduction in the mid-1970s when 3.6 million tons were grown has been reduced to around 2.2 million tons at present, while in the Mediterranean region production has increased by 50 per cent to around 2.5 million tons, almost exclusively of superior quality, seedless varieties.

Lemon and lime production has shown significant growth in the past decade and presently totals 6 million tons, most of which are lemons. The United States, Italy and Spain are the world's leading producers with 0.99, 0.88 and 0.65 million tons respectively. More than 60 per cent of the USA crop is processed into juice, compared with only 10 per cent in Spain. Mexico is the world's most important lime producer with around 0.6 million tons, most of which is sold fresh locally.

The world's production of grapefruit amounts to 3.8 million tons. The United States accounts for 2.3 million tons – 60 per cent of world output – of which more than half are processed. There is an increasing production of pigmented grapefruit in the USA and to a lesser extent elsewhere. Israel, Cuba, Cyprus and South Africa account for most of the remaining world grapefruit production, much of which is exported to Europe. Apart from Japan, where import restrictions were lifted in the mid-1970s and consumers have acquired a liking for imported grapefruit, the pummelo is preferred by most people throughout Asia.

Citrus Consumption

Since the early 1970s, consumption of fresh and processed citrus products (expressed in terms of fresh fruit equivalent) has risen from around 36 million tons to present-day output of 63 million tons. Much of the increase has been in processed form where over the past two decades the growth in citrus products has been at a rate of nearly 4.5 per cent per annum while the growth rate of fresh citrus consumption has averaged 3 per cent.

Although citrus consumption in developing countries has increased at a faster rate than in high-income countries, developed countries still account for around 65 per cent of the total market. The North American market alone accounts for nearly half the citrus consumption of developed countries, much in the form of juice. Western Europe accounts for a further third, 40 per cent of which is consumed in fresh fruit form but the growth in the citrus juice market has continued to increase at a very rapid rate in the past 15 years. In Japan annual consumption has been mainly as fresh fruit from local production since the import of fresh citrus and juice has been limited by quota.

Citrus consumption in the USSR and Eastern Europe has been no more than about 10 per cent that of Western Europe although the total population figures in the two areas are similar.

While the average world consumption of citrus is about 12 kg per person per annum, there is a wide variation between countries. Developed countries average 28 kg per person compared with only 6 kg in developing countries. In the United States consumption averages about 60 kg, while in Western Europe it is around 40 kg per person. This is in sharp contrast with average Eastern European and USSR consumption of less than 5 kg per person. Despite China's increased production to 2.6 million tons, a population of 1.2 billion means that average consumption is little more than 2 kg per person.

THE DEVELOPMENT OF NEW VARIETIES

There are six genera constituting the group of the true citrus fruit trees, three of which are of commercial importance: *Poncirus* (trifoliate orange), *Fortunella* (kumquat) and *Citrus*. Within *Citrus* there are eight important commercial species:

Sweet orange	(*Citrus sinensis*)
Mandarin	(*Citrus reticulata*)
Grapefruit	(*Citrus paradisi*)
Pummelo	(*Citrus grandis*)
Lemon	(*Citrus limon*)
Sour lime	(*Citrus aurantifolia*)
Citron	(*Citrus medica*)
Sour orange	(*Citrus aurantium*)

HYBRIDISATION

Most *Citrus* species are cross-fertile to varying degrees and *Poncirus* and *Fortunella* can also be hybridised with *Citrus*. It is widely believed that some *Citrus* species are the result of interspecific crosses; for example, the sweet orange is probably the result of a natural pummelo–mandarin cross and the grapefruit from a cross between pummelo and sweet orange. The lemon is a more complicated hybrid, possibly involving the lime, citron and perhaps pummelo.

Of the many citrus varieties in commercial production today, several are of hybrid origin. In some regions of the world, such as the Caribbean, China and Japan, important varieties have evolved from natural hybridisation events. The Temple, Ugli and Ortanique all evolved from chance crossings between different species in the Caribbean region. In Japan, several naturally occurring, interspecific hybrids are now extensively cultivated, such as Iyokan, Hassaku and Natsudaidai: varieties which are relatively unknown elsewhere. The important Chinese mandarin Tankan is almost certainly a natural tangor (mandarin [*tan*gerine] × *or*ange).

Despite the ease with which *Citrus* species can hybridise, man-made attempts to create new citrus varieties have, with a few minor exceptions, gone unrewarded. Several mandarin hybrids have achieved some status as distinct varieties such as the tangelos (mandarin [tangerine] × grapefruit) Minneola and Orlando, and the hybrid mandarins Nova and Fortune. As yet no major orange, grapefruit or lemon variety has resulted from controlled hybridisation. Of some considerable significance to citriculture has been the breeding of the intergeneric hybrids such as the citranges (*C. sinensis* × *P. trifoliata*) which have assumed great importance as rootstocks.

MUTATIONS

Citrus trees produce spontaneous mutations very readily. They are discovered as limb sports or bud sports and are also detected in nucellar seedlings. Navel oranges and grapefruit tend to produce more mutations than other citrus and many of today's leading varieties of these two citrus types occurred in this way. Numerous mandarin varieties have also arisen from spontaneous mutations. The wide range of Clementine varieties in Spain and most Unshiu (satsuma) mandarins in Japan arose from mutations discovered in the fairly recent past.

The speed at which citrus mutates can be increased by artificial induction using ionised radiation. While this technique has been used on only a limited scale, two highly pigmented grapefruit varieties, Star Ruby and Rio Red, have evolved from such programmes.

POLYEMBRYONY

One of the unusual features of *Citrus* is the development in one seed of two or more embryos and referred to as polyembryony. Although this feature is an impediment in any hybridisation programme it has been put to good use in other ways.

In most cases, the extra embryos develop

asexually from the egg sac or nucellus and are referred to as nucellar embryos. They are genetically identical to the mother plant (except when mutations have arisen in the nucellar embryo itself) since the genes of the pollen (male) parent are not involved in their formation. The initiation of nucellar embryos requires both pollination and fertilisation of the egg. It often happens that the sexual (zygotic) embryo does not develop, in which case almost all the embryos and subsequent seedlings are nucellar.

Not all *Citrus* species exhibit the characteristic of polyembryony and some, so far as can be determined, produce only zygotic embryos (pummelo, citron, Clementine, Temple and Persian lime, for example). Others produce only nucellar embryos (e.g. Sampson tangelo). Many lemon and lime varieties produce a significant percentage of zygotic seedlings but oranges, grapefruit and many mandarins usually have a low percentage. Nucellar seedlings have been of extreme value in the practically indefinite propagation of hybrids and in the production of clones which are usually free from most virus diseases carried by the mother plant. Many improved selections have been derived in this way but have not traditionally been given new variety names; instead they usually receive a code number but retain the original name.

The introduction of nucellar budlines is not straightforward, however, because all citrus seedlings and their immediate progeny have a long period of juvenility characterised by excessive thorniness, vigorous growth and slowness in coming into fruit production. Moreover, fruit quality is often inferior compared with old budlines. However, these shortcomings tend to disappear as both the trees and the budlines become older. The most valuable characteristics of nucellar selections are their initial freedom from viruses, a greater tree vigour and higher yields.

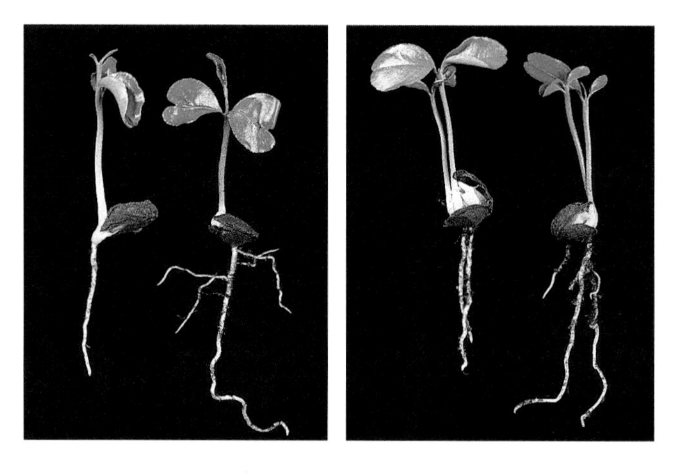

POLYEMBRYONY

Single seedlings from monoembryonic seeds (left). Polyembryonic seeds (right) each produce two (and sometimes more) seedlings

THE SWEET ORANGE

Citrus sinensis

The sweet orange, like most other citrus fruits, probably originated in the region between south-west China and north-east India and has been cultivated in southern China for several thousand years. There is speculation that the sweet orange is a natural hybrid of pummelo and mandarin but its true identity may never be known with certainty.

YOUNG AND ESTABLISHED ORCHARDS IN THE CENTRAL VALLEY, CALIFORNIA

Although some believe both sweet and sour oranges may have been known to the Romans there is little evidence they were ever cultivated by them. If they were, citriculture generally declined and became extinct in Europe when the Roman Empire fell.

In the Middle Ages, the introduction of the sour orange into the Mediterranean basin preceded that of the sweet orange, which was first cultivated in the region around 1450. This followed its spread across the Middle East from China by Arab traders and its introduction to the Ligurian coastal region by the Genoese traders. Important though the mid-fifteenth-century introductions were to the development of sweet orange growing, it was the Portuguese who were responsible for introducing from China the trees from which most of today's superior orange varieties originated.

Over the centuries of their cultivation many varieties have evolved to suit particular climatic conditions and the individual preferences of the local population. While many varieties, such as the Valencia, may best be described as 'dual purpose' oranges, being equally suitable for eating out-of-hand or for processing purposes, navel oranges are characterised by their unrivalled quality as fresh fruit but their unsuitability for processing, resulting from the development in the juice of an unacceptable flavour termed 'delayed bitterness'. Blood oranges have found favour in some Mediterranean areas, while acidless or sugar oranges are much appreciated in Latin American and Arabic-speaking countries but they are barely known in many other citrus-producing countries. A few other varieties are grown almost exclusively for processing and few enter into the fresh orange trade.

Total world production of oranges at 44 million tons is exceeded by that of grapes at approximately 64 million tons, but much of the grape crop is transformed into wine and other alcoholic beverages. In today's international fresh fruit trade the sale of 7.5 million tons of bananas exceeds that of oranges by 2 million tons. Nevertheless, when the consumption of fresh oranges in citrus-producing countries and the processing of juice is added to the total number of oranges entering world trade as fresh fruit, there is every justification in the claim that the orange is the world's favourite fruit.

It is convenient to place sweet oranges into four groups:

Navel oranges, of which the most widely grown variety is the Washington or Bahia.

Common oranges which include such well-known varieties as Valencia, Pera, Pineapple, and Shamouti.

Pigmented oranges, also known as blood, sanguina or sanguigna.

Acidless or sugar oranges, which have extremely low acidity and a rather insipid flavour.

MATURE NAVELINA TREE
Sagunto, Spain

NAVEL ORANGES

Navel oranges have the distinctive feature of having a small secondary fruit embedded in the apex of the primary fruit, and although this characteristic is sometimes found in other oranges and particularly in mandarins, it is never consistent and varies depending upon climatic factors.

Generally speaking, navels are the earliest maturing of orange varieties, producing seedless fruit of larger size than most others, with deep orange, easily peeled rinds, and a rich, sweet and pleasant flavour.

However, there are serious limitations to their production since the trees are less vigorous and less productive than those of many other varieties, and they are far more specific in their climatic adaptability. For example, navels thrive and produce superior quality fruit only in sub-tropical, Mediterranean-type climates and are unsuited to many regions where other orange varieties perform well. In contrast with Valencia oranges, which are grown in many citrus-producing environments and are one of the mainstays of production in tropical as well as semi- and sub-tropical regions, navels are far more restricted in their distribution. However, they are important in many countries worldwide and form a significant proportion of the citrus production of Spain, Morocco, Turkey, South Africa, California, Australia, Uruguay and Argentina.

While navels are rarely equalled, and never surpassed, by other oranges as a dessert fruit, they also have characteristics other than poor climatic adaptability which prevent their more widespread production. Although navel oranges yield less juice than most other oranges, it is the development of delayed bitterness in the juice which makes them unsuitable for processing. Unlike the bitterness in grapefruit caused by the compound naringin or in sour oranges by neohesperidin, bitterness in navel orange juice becomes evident only when the fruit is juiced and the bitter factor limonin is released from other closely related compounds. Although navel juice contains only extremely low levels of limonin, it is a very bitter compound which most people are able to detect at

NAVEL ORANGE

Navel oranges have a small rudimentary secondary fruit in the apex of the primary fruit

levels of no more than around five parts per million. For this reason navel orange juice cannot usually be used for the preparation of juice products unless it is first blended with juice of other varieties of low limonin content.

For some time it was widely believed that the navel orange we know today originated as a limb sport on a tree of the old-established Salata variety at Cabulla near Bahia (now Salvador), Brazil, some time prior to 1822. There is now much evidence to disprove this theory, for navel oranges are known to have grown in Spain and Portugal for many years prior to 1822 and it seems more likely that they were first brought to Portugal from China and thence to Brazil much earlier than this. However, the worldwide expansion in navel orange growing started only after the Bahia navel was sent in 1870 to the United States Department of Agriculture's facilities at Washington, DC, for propagation in glasshouses before being sent to California and Florida in 1873. It was from this importation that the Washington navel spread to other citrus areas.

Navels are genetically far more unstable than other leading orange varieties, with the result that countless selections have been made by growers and others in many parts of the world during the past century. Many have fruit characteristics which are almost indistinguishable from the Washington navel but a few have markedly different traits – particularly with respect to time of maturity. It is now possible in many navel-growing regions to extend the season from the normal two or three months to six months and sometimes longer. In California, for example, the harvesting season normally extends over a four- to five-month period with the Washington variety because of the several climatic zones in which navels are grown. California's recent import of the Australian variety Lane Late (see right) could lead to a harvesting season extending to perhaps eight months.

The navel varieties of note compared with the Washington are as follows:

BAHIANINHA

(Baianina Piracicaba)

This important Brazilian variety most probably originated as a bud mutation on a Bahia navel tree around 1907.

The Bahianinha tree is smaller and less vigorous than the Bahia (Washington) but more productive and less inclined to produce extremely large fruit. It is also less inclined to alternate-bearing and produces more consistently than the Washington. Bahianinha is climatically better adapted to hotter regions that experience heat-wave conditions during the fruit-set period.

Fruit size is medium to large, slightly more oval in shape than Washington and with a smoother, thinner rind; the navel is much smaller and often completely concealed. It is somewhat less easy to peel than Washington but not difficult. The flavour is richer, sweeter and has better acidity but is superior only when grown on a quality-inducing rootstock such as Troyer citrange.

Clearly better adapted to tropical and semi-tropical conditions than the Bahia (Washington), the Bahianinha has also performed well outside its native Brazil.

The disadvantage of not being immediately recognisable as a navel orange is offset by better fruit size, better quality and less susceptibility to post-harvest decay.

FISHER

(Fischer)

A popular early selection in California Fisher reaches the legal maturity standard earlier than Washington but not in colour break. It is reported to be as early as Newhall and of similar outstanding quality although the rind colour does not attain the same deep orange intensity.

GILLEMBERG

The Gillemberg is of unknown origin, possibly a limb sport of Palmer navel, and was discovered in 1985 at Gillemberg near Potgietersrus, Transvaal, South Africa, as a 30-year-old tree. The tree is more densely foliaged than Palmer and the leaves more tapered.

It is about six weeks later maturing than Palmer and retains its quality well when left on the tree for prolonged periods. The fruit is smaller than Palmer but otherwise similar and has outstanding flavour even when grown on rough lemon rootstock in the cooler citrus-producing areas of South Africa. It seems to have outstanding potential elsewhere.

LANE LATE

Lane Late was discovered in 1950 as a bud sport on a Washington navel tree on the property of L. Lane, Curlwaa, near Mildura, Victoria, Australia.

Tree characteristics are indistinguishable from Washington, and so too are many physical features of the fruit, except rind texture which tends to be somewhat smoother, and the navel which is smaller and less protruding. A significant characteristic of the Lane Late is the lower limonin content of its juice. However, it is appreciably later maturing and colours up much more slowly by as much as four to six weeks in some localities. When Washington navels reach peak maturity in Mildura around July, Lane Late has lower sugars but similar acidity, and the flesh is still somewhat immature.

In Australia harvesting does not commence until September – about the same time Valencia oranges reach the minimum maturity standard – by which time Lane Late has developed a high sugar content and fine rich flavour with a good acid level.

It is possible to hold the fruit on the tree for an incredibly long period: until autumn (i.e. April in Australia), particularly when grown on trifoliate orange rootstock. In California the rind regreens in late spring and summer.

Extensively grown in inland districts of Australia where a reported 50 per cent of navel plantings are Lane Late, it is now being evaluated elsewhere. In South Africa it has proved very disappointing for, despite its excellent flavour, it develops only yellowish-orange rind colour, is oily to peel, with much albedo remaining on the segments which are tough and excessively raggy, and is low in juice.

LENG

The Leng navel originated as a limb sport of Washington on the property of the Leng brothers, Irymple, near Mildura, Australia, and was discovered in 1935.

Reports vary regarding tree vigour and size, but from my own observations its characteristics are little different from Washington, except for its much narrower leaf shape.

Fruit size is markedly smaller than Washington and regarded as being of medium size, while rind texture is much smoother and appreciably thinner but more prone to splitting near the navel. Rind colour is as good and often better than Washington, developing into a deep sometimes reddish-orange intensity.

The flesh is very juicy and soft in texture with good flavour and deep orange colour. Maturity is 10 to 14 days earlier than Washington, and the fruit holds well on the tree without deteriorating in quality.

Leng has been extensively planted in Australia's arid inland citrus-producing areas along the Murray and Murrumbidgee Rivers. In California a recent evaluation suggests it performed disappointingly, due primarily to small fruit size and light crops although fruit quality was excellent. In South Africa productivity is good, but much of the fruit is too small for export.

NAVELATE

Discovered in 1948 by D. A. Gil at Vinaroz, Castellón Province, Spain, as a limb mutation on a Washington tree, it was released for propagation in 1957.

The trees are vigorous and slightly larger in size than Washington but are thorny.

The fruit is medium to large in size, somewhat smaller than Washington, and has a smaller and often concealed navel. The rind is of similar texture but is thinner, more leathery and somewhat less easily peeled. In Spain external colour develops several weeks later, while internally it matures at about the same time. However, it can be left to hang on the tree for four months or more without appreciable loss of quality.

The Navelate has always promised to be a significant breakthrough in navel orange production. However, today it accounts for no more than 5 per cent of Spain's navel production, although it presently accounts for around 10 per cent of navel orange plantings.

It is an erratic performer even in Spain where light cropping has been a feature, but this is now believed to have been due to inadequate irrigation at flowering time. Elsewhere Navelate has proved inferior. For example, in Morocco it has lower juice than Washington and is reported to mature not significantly later, while in South Africa it produces light crops of fruit which are often extremely difficult and oily to peel.

NAVELINA

(Dalmau)

Originating in California in 1910 and first named Smith's Early navel, this variety was released by the University of California, Riverside, to the IVIA Research Institute, Valencia, Spain, in 1933 where it was later renamed Navelina; but it was not released to nurseries until 1968.

The tree has good vigour but never attains more than small to medium size when mature. Nevertheless, it is productive and the fruit matures at least two weeks earlier than Washington, reaching acceptable minimum internal quality standards in Spain in mid-October while still having inadequate colour. However, it may be degreened without problems and achieves satisfactory colour as a result.

Navelina fruit size is smaller than Washington and slightly more oval in shape, especially at the navel-end where the navel itself is smaller and more concealed. Rind texture is very slightly smoother, but of the same thickness as Washington, and it develops similarly good colour when fully mature.

Always of good quality, Navelina is generally agreed to fall somewhat short of the excellent flavour of Washington. However, it is currently being planted on a far greater scale in Spain and, with Newhall navel, it accounts for over 50 per cent of orange plantings.

No distinction is made between Navelina and Newhall when the crop is marketed: they are both sold as Navelina and jointly account for over 55 per cent of the 1.5 million tons of Spanish navel production.

Although interest has been shown in other countries, for example in South Africa, where it constitutes almost 20 per cent of navel plantings, Navelina is still essentially a Spanish variety.

NEWHALL

The Newhall is of Californian origin, having been discovered as a limb sport of Washington in the Duarte area and propagated by P. Hackney of the Newhall Land Co, Piru.

Tree and fruit characteristics under Spanish conditions are indistinguishable from Navelina, apart from fruit maturity, which is very slightly advanced due to somewhat lower acidity. In California. Newhall develops much deeper early rind colour than the Washington. It is equally as popular as Navelina with Spanish producers. The two varieties are marketed together as Navelina but neither is grown elsewhere to any extent.

PALMER

The standard navel variety in southern Africa, of nucellar origin from Washington, Palmer is indistinguishable except for greater tree vigour and freedom from exocortis viroid.

SKAGGS BONANZA

A Californian variety of fairly recent introduction, it originated as a bud mutation of Washington.

Skaggs Bonanza trees have similar characteristics to Washington but are smaller, denser, more productive and come into bearing at an earlier age. Fruit size is medium to large but otherwise like Washington, except for colour development and maturity which is earlier by about two weeks. Even at full maturity, however, it does not develop the same deep orange intensity as the Washington. It has good sweet flavour, but Skaggs Bonanza does not hang well on the tree. It drops the earliest of all navel strains and, if not harvested early in the season, is mostly on the ground and lost.

Gaining in commercial importance in California, Skaggs Bonanza is being evaluated elsewhere and has achieved some significance in Chile.

THOMSON

One of the earliest selections to be made of the Washington, Thomson originated as a bud sport on the property of A. C. Thomson at Duarte, California, about 1891. In its early years it was regarded as having potential importance because of its attractive appearance and maturity 14 days in advance of Washington. It lacked tree vigour and size, often being semi-dwarf compared with Washington. Fruit size was comparable, but rind colour and yield were somewhat inferior and, more importantly, quality was much inferior. Although extensively planted in other citrus-producing countries, few plantings remain today except in Chile where it is one of the principal varieties, but where the climate is regarded as being marginal for the production of high quality oranges.

WASHINGTON OR BAHIA

Washington navel trees are of medium vigour and size, somewhat drooping in habit and with poor adaptability to extreme climatic conditions such as hot dry weather at flowering.

The fruit matures early and is large, round to slightly oval in shape with a well-developed navel which is sometimes concealed but usually exposed, and occasionally protruding. The rind has a slightly pebbly texture, is moderately thick and fairly brittle but peels easily. When grown in sub-tropical Mediterranean climates rind colour is deep orange and the best of all non-blood varieties, but in Brazil where it is produced in significant quantities the hot, humid, semi-tropical conditions lead to very poor colour development.

The segments are easily parted, the flesh firm, crisp but tender, moderately juicy and of fine, rich, sweet flavour but with adequate acidity.

The fruit holds well on the tree except when grown under unfavourable conditions or on too vigorous a rootstock resulting sometimes in the development of granulation, which sometimes continues to progress after harvest.

WASHINGTON OR BAHIA

COMMON ORANGES

These were often referred to in the past as blond or white varieties (blanca in Spain, biondo in Italy) to distinguish them from pigmented and navel varieties of sweet orange. Common oranges form a large and diverse group with a wide range of tree growth and fruit quality characteristics.

The more seedy varieties are derived from seedlings and are collectively referred to as Comuna in Spain (Comune in Italian) and Beladi or Maladi in Arabic-speaking countries. As seediness is considered an important disadvantage on the fresh fruit markets, especially throughout Europe, the cultivation of these varieties has been greatly reduced in the past two or three decades in favour of less seedy or completely seedless types, especially navels.

However, for processing purposes, seediness is of little or no importance and instead emphasis is placed on the matters of fruit yield and sugar content (or solids) – what is termed in the United States 'pounds solids per acre'. Other factors such as juice colour, freedom from delayed bitterness and time of maturity are also of considerable significance with orange varieties for processing.

The more important common orange varieties, and those of potential importance for both processing and for eating out-of-hand are as follows.

AMBERSWEET

Released in 1989, Ambersweet is a hybrid orange resulting from a cross between a Clementine × Orlando tangelo hybrid with a seedling midseason orange made in 1963 by C. J. Hearn and P. C. Reece, USDA, Orlando, Florida.

Ambersweet trees are upright in shape, densely foliated and moderately vigorous, with thorns on the young shoots. The trees are moderately cold-hardy.

Fruit is of medium size and of distinctive shape, being rounded at the apex and tapering towards the stem-end, and having a somewhat pyriform overall shape. The stem-end is sometimes

Ambersweet

Berna or Verna

furrowed and occasionally fruit may have a small navel.

The rind is very firm, moderately thick but peeled with utmost ease leaving the segments practically free of albedo, although it is rather oily. Of good colour, Ambersweet is mature in mid-October and can be harvested until the end of December under Florida growing conditions. Ambersweet flesh is very tender and rag-free, while the juice has very good flavour slightly resembling that of the Clementine and its deep orange colour is far better than the Hamlin harvested at the same time.

Excessively seedy in mixed variety plantings but virtually seedless when grown in solid blocks, it is a reliable bearer and does not require cross-pollination for fruit set.

Ambersweet has potential that may be somewhat limited as a fresh fruit due to its appearance, although external quality cannot be properly evaluated until it is grown in other climates. It is clearly worthy of evaluation on account of its fine juice characteristics for such an early orange variety.

Belladonna

This is an old Italian variety of unknown origin which is still cultivated on a considerable scale throughout the mainland areas of Italy but is no longer being planted to any extent.

Maturing in December and hanging well on the tree for several months, the fruit is of medium size, oval in shape with a medium to thickish rind. The flesh colour is deep orange but the quality is only fair having very little richness and only moderate sugar levels. It is mostly used for processing.

Berna or Verna

The Berna (or Verna) is a late maturing Spanish variety of unknown origin, roundish-oval to oval in shape, and small to medium in size. The rind has a pebbly texture and is of good orange colour but will regreen if left on the tree too long, i.e. after May. The rind is fairly thick and the juice content only moderate, but most fruits are seedless.

The flavour, although moderately sweet, lacks richness. Because its quality and size cannot match those of other late varieties, such as Navelate and Valencia, and it will not hang nearly so long on the tree as the latter, its importance is waning. It is rarely planted today, although there are still about 3,000 ha in production in Spain. It is not cultivated elsewhere on any scale.

CADENERA

Cadenera most probably originated from the Comuna, being discovered about 1870. The trees are vigorous and of medium size, while the fruit is medium to large, almost round in shape and with a fairly thin and very slightly pebbled rind texture. Maturing in November, it has a high juice content, is virtually seedless, and has a good, fairly rich flavour.

Cadenera was once the most popular of Spanish oranges but has gradually lost ground to navel varieties and is currently grown on only around 1,500 ha.

CASTELLANA

A Spanish variety of unknown origin with trees that are vigorous and productive with medium to large size fruit, which is fairly seedy, typically eight to ten seeds per fruit. The smooth rind is thin and easily peeled, while the flesh has a good juice content of fine quality.

Harvested from November onwards, the fruit hangs well on the tree without marked loss of quality. It is grown principally in the Almeria region of Spain, but its seediness is a serious limitation to its continued cultivation.

DELTA SEEDLESS

Originating in 1952 near Pretoria, Transvaal, South Africa, Delta Seedless is more productive than the Valencia, with slightly larger fruit. More importantly, it bears much of the fruit inside the tree canopy, resulting in its being less susceptible to wind blemish and consequently having a better packing-out percentage. The fruit is of good quality, nearly seedless, and only a small minority of fruits have one or two seeds. It has a lower sugar content and acidity than the standard Valencia, especially in early production years, and reaches acceptable maturity one to three weeks earlier. The Delta Seedless should preferably be grown on a high quality rootstock.

HAMLIN

Arising as a chance seedling in 1879 near De Land, Florida, and named by the owner A. G. Hamlin a few years later, the Hamlin was developed on a large scale in its native Florida and to a limited extent elsewhere worldwide.

The tree is moderately vigorous, developing medium to large size when fully grown, and it is one of the more cold-tolerant orange varieties. It is productive but fruit size is small, sometimes too small for the fresh fruit market. It develops early colour, maturing at almost the same time as Washington navel. This earliness is a decided advantage where there is a threat of winter freezes – in Florida, for example, where it is the leading early orange variety.

The fruit has a tendency to split and in Brazil it drops from the tree if left on too long. In semi-tropical areas such as Brazil and Florida, the rind is smooth, thin and pale orange in colour, whereas in Morocco and Turkey peel thickness is moderate with a slightly pebbly texture, and colour develops well to deep orange.

Hamlin fruit is not difficult to peel, the flavour is sweet but not especially rich and juice content is high, with only an occasional seed. The juice of

HAMLIN

much of Florida's Hamlin crop is pale in colour and has a high limonin content, particularly early in the season. Poor colour and flavour necessitate its being blended with juice of better quality, later maturing oranges.

Hamlins make up about 5 per cent of Brazil's production, much of which is sold fresh. The Hamlin from Morocco is of good quality but there, as in South Africa, its small size precludes its continued long-term production, especially when in competition with better quality navel oranges of the preferred sizes.

An interest in the Hamlin is being shown in China where it might well succeed, as the climate in many of the citrus-producing areas is not unlike that in Florida and Brazil. Moreover, there is a marked preference among Chinese consumers for oranges with low acidity and sweet, albeit somewhat insipid, flavour.

MARRS

(Marrs Early)

This variety was discovered as a limb sport on a navel tree on O. F. Marrs' property at Donna, Texas, in 1927.

The trees are small but moderately vigorous and produce heavy crops from an early age.

The fruit is medium to large in size, round in shape and without a navel. It is borne mainly in clusters on the outside of the tree. The rind is smooth, thin, light orange in colour and fairly easily peeled. Because of its low acidity it reaches legal maturity in early September, well ahead of other varieties. At this time the quality is only fair, being sweet but somewhat insipid. However, the flavour improves appreciably if the fruit is left to mature fully in November.

The flesh is normally pale in colour and has a slightly lower juice content than Hamlin. It is moderately seedy, commonly having seven to ten seeds per fruit.

Grown almost exclusively in the Lower Rio Grande Valley in Texas on perhaps no more than 2,000 ha, Marrs is clearly better suited to the local climatic conditions than other early varieties such as navels which produce very light crops.

MIDKNIGHT

Of unknown origin, this variety was first noticed as a slightly earlier maturing tree growing in a Valencia orchard at Addo, Cape Province, South Africa, by A. P. Knight (*mid*season + *Knight*) about 1927 but its potential was not recognised until the 1970s.

Its main improved characteristics are the exceptionally high juice percentage, better flavour, near seedlessness and significantly larger fruit size. Midknight matures two to four weeks earlier than the Valencia but holds on the tree just as late. It is, however, more difficult to peel because the rind is thinner and more tightly adhering, and often oily. It appears to be best suited to climatic conditions in which navels produce their finest quality.

NATAL

The Natal variety is of unknown Brazilian origin, resembling the Valencia in most tree and fruit characteristics but maturing even later and, as the name suggests, usually harvested around Christmas in Brazil.

Rather surprisingly, this late maturing characteristic has seldom been evaluated in other citrus-growing areas worldwide, with the result that Natal is grown only in Brazil. However, there it is a most important variety for the processing industry, second only in importance to the Pera and ahead of the Valencia in Sao Paulo State, with 15 per cent of all orange production.

OVALE

(Calabrese Ovale)

Like so many Italian varieties, the Ovale's origin is obscure but goes back many years.

The tree has some good characteristics: it is vigorous, attains large size, and is productive under favourable conditions. However, it is adversely affected by extremes of temperature, slow to come into bearing and has a tendency to produce out-of-season fruit.

The fruit is of only medium size and has a smooth, well-coloured rind which is fairly thin and easily removed despite its tight adhesion. Juice content is good, as is its colour. Flavour is sweet and fairly rich and there is a high sugar level. The fruit is sometimes seedless, although a significant proportion have just one or two seeds per fruit.

Ovale is the latest maturing Italian orange variety, and is similar to the Verna in Spain, both varieties maturing a few weeks earlier than Valencia. The fruit hangs well on the tree before losing quality, but not to the same degree as the Valencia which may remain unpicked and in good condition for months after reaching maturity, given favourable climatic conditions.

Most Ovale production is for the Italian fresh fruit market and is sold out of cold storage throughout the summer period.

PARSON BROWN

PARSON BROWN

This variety originated in 1856 as a seedling at the home of Reverend N. L. Brown near Webster, Florida, from fruit brought to Savannah, Georgia, from China. In 1874 budwood from the parent tree was propagated and the variety name given was 'Parson Brown'. Until the 1920s this was the leading early orange variety in Florida before being replaced by Hamlin, which is somewhat larger and far less seedy: Parson Brown regularly has an average of 15 seeds per fruit.

The tree is vigorous, large in size and very productive. The fruit is round in shape with a smooth, fairly thin, well-coloured rind. It matures early in October and November in Florida. The flesh and juice are not well coloured compared with later varieties such as Pineapple and Valencia. However, it is sweet and well flavoured and the juice content is high.

Parson Brown's popularity has diminished gradually in Florida. It never achieved any significant importance in any other citrus industry as a 'dual purpose' variety, as did Hamlin, because of its seediness and fruit size.

PERA

The origin of this important Brazilian variety is unknown but believed by some authorities to be the same as Lamb's Summer of Florida, a variety regarded there as being virtually indistinguishable from Valencia but which never attained much significance.

Pera trees are vigorous, upright in habit, large and very productive. Fruit is smaller than the Valencia and ripens earlier, being regarded as a late midseason. It is harvested from June to October in Sao Paulo State but some fruit is often exported several weeks earlier before attaining full maturity.

Slightly oval in shape, with a smooth thin rind that adheres fairly tightly but is not too difficult to peel, Pera is pale orange both externally and internally, although this colour does improve later in the season. Pera fruit does not have the same sugar content as Valencia and the acidity is usually low, making for a somewhat sweet but insipid flavour lacking in richness. Like the Valencia grown in Sao Paulo State, it has a high juice content but when eaten it is rather tough and raggy. It is moderately seedy with five to ten seeds per fruit. However, on account of its exceptional productivity, Pera is well suited to the processing industry of Brazil.

PERA
Brazil's most important orange variety

The Pera is inclined to uneven flowering, which leads to fruit of differing maturity being borne on the tree at the same time, a condition further aggravated when it is grown without irrigation.

An appreciation of the importance of the Pera in Brazil may be gained from the fact that well over 50 per cent of Brazil's massive 15 million ton orange crop is derived from this variety. It is not grown outside Brazil.

PINEAPPLE

This is thought to have its origin in fruit brought to Charleston, South Carolina, USA, from China and planted by Reverend J. B. Owens at Sparr, near Citra, Florida, around 1860. Budwood from one of these trees, which had already been named 'the Pineapple tree', was eventually propagated as the Pineapple variety. Some people maintained the tree was shaped like a pineapple, others said the fruit resembled a pineapple in either smell or flavour.

The Pineapple tree has moderate vigour, medium to large size and is very productive. However, it has a tendency to alternate-bearing, and when carrying a heavy crop during mid-winter the tree is particularly susceptible to cold damage. The fruit is also very susceptible to pre-harvest drop. Also, occasionally the peel disorder, Pineapple-pitting, occurs on heavy bearing crops.

The fruit is medium in size, almost round in shape, and has a fairly smooth, thin rind which attains good colour – better than Parson Brown – when fully mature. In Florida it is at its best in January and February. The juice content is high, with good colour and a fine, very sweet and rich flavour. However, it is even more seedy than Parson Brown, regularly having 20 or more seeds per fruit.

Grown in Florida primarily for its fine juice processing characteristics, Pineapple was also planted to a very small extent in South Africa and Brazil for the fresh fruit export trade, but the excessive seediness brought about its discontinuance several years ago.

SALUSTIANA

Originating as a bud mutation from the Comuna and propagated by Don Salustiano Pallas near Enovo, Valencia Province, Spain, in 1950, this variety is now the second most important blanca variety in Spain (after Valencia), occupying about

PINEAPPLE

SALUSTIANA

SHAMOUTI

4,000 ha and accounting for around 6 per cent of all Spanish orange trees currently being planted.

Salustiana trees are vigorous, well developed and very productive. The fruit matures in December but is not at its best until late January, and hangs well on the tree until late April: about the time the Valencia reaches legal maturity in Spain.

The fruit is medium to large in size, with a finely pebbled and medium thick rind, while the flesh is especially tender, very smooth in appearance and quite distinct in this regard. Virtually seedless and with a fine, rich and sweet flavour, it is the outstanding midseason blanca orange variety in Spain.

It is also produced to a limited extent in Morocco, where the quantity rivals that from Spain. Under summer rainfall conditions at Nelspruit, South Africa, Salustiana has many true-to-type characteristics but acidity is low and flavour only fair and lacking richness.

SHAMOUTI

(Jaffa, Cyprus Oval)

The Shamouti originated as a bud mutation on a local Beladi or common orange tree in 1844 in an orchard near Jaffa, Palestine (now Israel). Several strains have been identified: some earlier, others later maturing than the common type, and some pigmented or seedy.

The Shamouti tree has a distinctive upright growth habit, is moderately vigorous, thornless and has large broad leaves. Its productivity and external and internal fruit quality are greatly affected by climatic conditions, rootstock and soil factors. In Israel the best quality Shamouti is produced in the narrow coastal belt and on Palestine sweet lime rootstock, while in Cyprus it produces fruit of similar quality when propagated on sour orange.

Shamouti fruit has distinctive characteristics which might at first sight indicate the fruit is not of superior quality. It is medium to large in size and oval in shape with a slightly flattened stem-end. The rind texture, particularly at the stem-end, is pebbly (somewhat rough on larger fruit) and fairly thick. One of the easiest oranges to peel without releasing much rind oil, the Shamouti has a distinctive fragrance and unique outstanding flavour, being rich and sweet but with adequate acid to give a good sugar to acid ratio.

The segments are particularly tender and the fruit is virtually seedless. Unfortunately the juice yield on a weight basis is rather low; moreover, Shamouti juice develops delayed bitterness like navel oranges which discounts its value compared with that from other varieties such as Valencia.

A midseason variety reaching acceptable internal quality in mid-December in Israel but at peak maturity a few weeks later, the Shamouti develops good colour and hangs well on the tree without its quality deteriorating. If picked when fully mature it stores particularly well.

Although grown mainly in Israel and Cyprus, excellent Shamoutis are also produced in the Mersin region of Turkey and in parts of Swaziland. Despite its fine quality characteristics the volume produced in the eastern Mediterranean region is gradually declining due to strong competition, notably from navel oranges and mandarins.

TROVITA

Most probably a seedling of Washington navel found in California at the Citrus Research Center, Riverside, around 1916 (the name meaning 'discovered' in Esperanto) and released in 1935.

The tree is vigorous, upright in habit and productive, with a slight tendency to alternate-bearing.

Trovita fruit is roundish, slightly flattened at the ends and is early maturing although somewhat later than the Washington navel. In Israel the fruit is still slightly green in early December, but later it develops good colour. Fruit size is smaller than Washington navel but still medium to large.

The rind is smooth, very thin and very easily and cleanly peeled, with little albedo adhering to the segments. The juice content is moderate, but somewhat better than Washington navel, the flesh is tender and it is fairly seedy with six or sometimes up to ten seeds per fruit. With its sweet flavour the quality in early December is comparable to the Spanish Navelina.

It ships far better than the Washington navel grown under the same conditions, but has the disadvantage of having no navel or other features to make it distinguishable from other varieties of common orange.

Probably well suited to semi-desert conditions where the Washington navel does not produce good quality fruit, Trovita has only been planted in Israel and there its future is uncertain.

VALENCIA

(Valencia Late)

It is commonly assumed, perhaps understandably considering the name, that the Valencia is of Spanish origin. However, the variety first became of interest in the Azores and is almost certainly of old Portuguese origin.

No other citrus variety has a more fascinating and improbable history than that of the Valencia, which is now the world's most important orange. Sent from the Azores in the early 1860s to Thomas Rivers, a nurseryman at Sawbridgeworth, England, it was first named Excelsior. Rivers had recognised its late maturing characteristic and believed it suitable for growing in containers in the fashionable orangeries of country houses.

He sent trees of the Excelsior and other varieties to S. B. Parsons, Long Island, USA, in 1870, who in turn supplied both A. B. Chapman of San Gabriel, California, in 1876 and F. H. Hart, Federal Point, Florida, the following year. Chapman named the variety Rivers Late, while Hart's trees were initially designated as Hart's Tardif (or Hart Late).

In 1887 Rivers Late was renamed Valencia Late at the suggestion of a Spanish citrus expert visiting California who believed that it bore great similar-

TROVITA

VALENCIA
A well-coloured variety when produced in a sub-tropical climate

VALENCIA
In tropical regions such as Cuba the rind is poorly coloured and the flesh and juice are pale orange

ity to a late maturing orange grown in the Valencia region. It was a decade or so later that authorities recognised that the Hart Late and Valencia Late were in fact the same variety.

Its outstanding qualities were soon recognised and the Valencia orange was to change the face of citrus production on a world scale so that today it is the leading variety in many citrus-producing countries and an important one in others. There is no other citrus variety more widely grown and on such an extensive scale. The Valencia leads production in Argentina, Australia, California, Florida, Morocco, southern Africa, Uruguay and other countries and is an important variety in Brazil, Israel and elsewhere. Somewhat surprisingly, it has not been extensively planted in Spain, but Valencia production is now increasing and has replaced the Berna as the principal late maturing variety, although it still accounts for only 8 per cent of the country's annual orange crop.

Valencia trees are vigorous, upright, large and very prolific but have some tendency to alternate-bearing.

Fruit size is medium to large, roundish-oblong in shape, with a well-coloured, moderately thin rind of smooth, but sometimes finely pebbled, texture. Valencia rind is prone to creasing particularly on some rootstocks. Not difficult to peel when fully mature, the rind is thin and leathery and the flesh well coloured, with a very high juice content of outstanding colour and good flavour although sometimes slightly acidic except when fully mature. Seeds typically number two to four per fruit. It is the latest maturing of all sweet orange varieties (with the exception of Natal in Brazil) and often hangs late into the summer of the following season without losing quality except that the rind may regreen somewhat while still on the tree. Moreover, the later it is picked, the smaller the next year's crop because of the 'two crops on the tree at one time' phenomenon.

In tropical regions, the rind, like that of other citrus varieties, never attains good colour and is often greenish, extremely thin and tightly adhering while the flesh and juice are a paler orange than that of Valencia fruit produced in sub-tropical Mediterranean climates. Valencia juice has excellent processing characteristics, including a deep orange colour, and the fruit ships and stores exceptionally well.

There are several clones of the Valencia, some of which have been given separate variety names. The most common improved selections are all thought to be nucellar in origin, having the following characteristics.

Olinda Originating as a seedling on O. Smith's property at Olinda, California, in 1939 and released in 1957, it is indistinguishable from Frost except the fruit has a tendency to produce more chimeras, i.e. fruit which is ribbed or has variegated rind.

Cutter Trees are more vigorous, larger and more thorny than the standard Valencia as well as being somewhat slower to come into bearing but are more productive. The fruit is indistinguishable from other Valencias but has the distinct advantage of producing fewer fruit chimeras.

Frost Grown as a nucellar seeding in 1915 by H. B. Frost at the Citrus Research Center, Riverside, California, and released in 1952, this selection has exceptional vigour and is highly productive, but fruit quality is very similar to old budline Valencia.

Other selections have been evaluated but have not gained more than local popularity – Armstrong and Campbell in California, Lue Gim Gong in Florida, Ksiri in Morocco and Harwood in New Zealand – but two South African clones, Delta Seedless and Midknight (see pages 17 and 18), have recently attracted attention and are now regarded as separate and distinct varieties.

WESTIN

Of unknown origin, the Westin is a Brazilian variety formerly referred to as Clementina but was renamed in honour of Professor Philippe Westin Cabral de Vasconcellos.

The trees are large, vigorous and very productive. The fruit is of medium to small size, virtually seedless, with a high juice content often exceeding 60 per cent, and of good colour. The flavour is rich with a good acid level.

A midseason variety, maturing in June and July in Sao Paulo State, it is reputed not to hang well on the tree thereafter. It is particularly well suited to processing on account of its well-coloured juice, but it is outyielded by Pera and therefore less favoured by producers.

PIGMENTED ORANGES

Pigmented or 'blood' oranges were thought to have originated in the Mediterranean area, probably on Malta or Sicily, but it has recently been suggested that they, like navel oranges, are indigenous to China.

Blood oranges have many similarities with common (blond) oranges but differ when grown under conditions which favour the synthesis of red pigments in the flesh and juice and sometimes on the rind. These red pigments, known as anthocyanins, are a common occurrence in many plant tissues, particularly in leaves and flowers. In blood oranges they develop only when there are low night temperatures and it is not until the autumn and winter that the fruit develops its familiar reddish characteristics. In tropical and semi-tropical climates blood orange varieties never develop any degree of pigmentation, although occasionally when cold stored, fruit will develop, post-harvest, the slightest amount of blood flecks in the flesh.

It should be noted that the pigments in red and pink grapefruit are lycopene and carotene and are not due to anthocyanins as they are in blood oranges. In addition, the climatic conditions required for good colour development of pigmented grapefruit are high temperatures during fruit growth whereas with blood oranges chilling is an important prerequisite.

Marked variations are often found in the extent of rind pigmentation of blood oranges, the trend being towards better colour developing on shaded than on exposed fruit. Moreover, rind colour development is often better on the shaded side of individual fruits.

Similarly, variations in flesh and juice colour are very considerable even from fruit produced in the same cluster. Pigmentation is invariably most pronounced towards the stylar-end of the fruit while it is commonly restricted in most varieties to the flesh immediately adjacent to the segment walls. Blood oranges often have a distinctive flavour sometimes described as resembling that of raspberries or cherries and in most but not all varieties this seems to be most pronounced in fruit with the strongest flesh pigmentation.

Blood oranges have been cultivated in the Mediterranean region for several centuries, particularly in the western Mediterranean countries: Italy (known as Sanguigna and Sanguinella), Spain (Sanguina), Morocco, Algeria and Tunisia (Sanguine). Until recently blood oranges were particularly popular in Europe but production limitations, particularly the preponderance of small size fruit, and the development of improved navel selections, have meant that blood oranges, like seedy common oranges, have gradually lost favour on the export markets. Although some varieties of bloods will probably never attain their former popularity, there are others of such outstanding quality, including the Tunisian Maltaise and Italian Tarocco, which are among the world's finest dessert oranges and are well worthy of evaluation in other countries within and without the Mediterranean region.

Outside the Mediterranean basin, blood oranges are grown to only a limited extent and form a negligible percentage of production in most citrus-growing countries.

Traditionally blood oranges have not been considered suitable for processing since the pigments tended to deteriorate to a brownish muddy colour. However, in Italy fine blood orange juice is produced, more closely resembling tomato juice than the familiar product, and this retains much of its distinctive flavour and aroma. Freshly squeezed blood orange juice makes an attractive refreshingly different drink from its more traditional counterpart.

DOBLE FINA

(Oval Sangre, Sanguinia Oval)

Of unknown Spanish origin, Doble Fina trees are vigorous and productive, but the fruit has several important shortcomings which have led to its rapid decline in popularity from being the principal blood orange on the European markets.

The trees are small to medium in size and the leaves are sparsely distributed over the open habit.

DOBLE FINA

Medium in size, Doble Fina fruit shows great variability in shape from round to oval, with and without a concealed navel, while external pigmentation varies from intense to non-existent. In this respect it is almost as variable as the Italian Moro variety. Rind pigmentation correlates well with internal colour, the more intensely blushed fruit having deeper flesh pigmentation.

The rind is smooth, medium to thick, tightly adhering but not difficult to peel. Juice content is fairly low although the flesh is tender, and there is always the distinctive blood orange flavour irrespective of the degree of pigmentation. Of only moderate sweetness and slightly acidic, the flavour is not unlike that of Moro but does not have the same marked acidity. It is moderately seedy, averaging about five seeds per fruit.

Harvested in January in Spain, the Doble Fina does not hold well on the tree and drops easily. It is the parent of the other blood varieties Doublefine Ameliorée (or Washington Sanguine), Sanguinelli and Entrefina.

MALTAISE SANGUINE

*(Maltaise de Tunisie,
Maltaise Semi-sanguine,
Tunisian Maltaise)*

An important, high quality semi-blood variety, the Maltaise Sanguine is of unknown origin, but Malta is most probable. It is grown extensively in Tunisia and to a lesser extent in Morocco.

The tree is moderately vigorous but attains only medium size when mature and is of only average productivity with a tendency towards alternate-bearing.

Fruit is of medium size, slightly oval in shape, with a finely pebbled rind texture. External colour is predominantly orange but occasionally has the slightest of blushes. The rind often feels soft, is of medium thickness and is easily peeled.

The flesh is well coloured but never more than moderately well pigmented and often only very slightly so. It is, however, very tender and rag-free (like a Shamouti), extremely juicy and virtually seedless.

The Tunisian Maltaise has outstanding flavour which is regarded by many, including myself, as being the finest quality of any non-navel orange; in France it is spoken of as the 'Queen of Oranges'. It is very sweet but with adequate acidity and has a particularly delicate flavour which, when combined with its tenderness, seedlessness and high juice content, forms the near ideal dessert fruit.

MALTAISE SANGUINE
'Queen of Oranges'

Maturing in late January and February, the fruit hangs only moderately well on the tree before losing quality but may be stored for a reasonable period post-harvest without significant loss in quality or condition.

Most of Tunisia's production is in the Soliman district of Cape Bon to the south-east of Tunis and totals around 100,000 tons annually, much of which is exported to France where its fine quality consistently earns significant premiums over the best Spanish navel oranges.

Although probably very specific in its climatic requirements to fully attain its best quality, and despite producing fruit of only medium size, the Maltaise Sanguine has such outstanding flavour that it deserves to be evaluated on a broader scale under conditions similar to those which exist in Tunisia.

In addition to the standard Maltaise Sanguine, two derivatives have been identified:

Maltaise Sanguine Bokhodza An early maturing, low acid selection which is reportedly more decay-prone and does not ship well.

Maltaise Tardive Barlerin A late maturing type but otherwise like the standard Maltaise Sanguine.

MORO

Of recent origin from the Sanguinello Moscato variety, the Moro was extensively planted in the 1950s and 1960s in Italy, and particularly in eastern Sicily where it is now the predominant variety.

The tree has moderate vigour, eventually reaches medium size and is very productive. The fruit is medium to medium-small in size but it is so variable in size, shape, rind texture and colour that there is no such fruit as a 'typical' Moro. Round, oval or distinctly oblate in shape, with or without a

MORO
This exhibits great variability in internal as well as external characteristics

navel, rough or smooth textured: all possible combinations are found in fruit on the same tree.

Some may have rind heavily or moderately blushed, others 'blond', thin and thick, tightly and loosely adhering. Like most blood oranges, internal colour depends greatly on seasonal climatic factors but no variety has the complete range from 'blond' to burgundy-coloured flesh like the Moro. The Moro selection planted in California on a limited scale is consistently well coloured both externally and internally.

Fruit may be moderately seedy or seedless and is usually easy to peel. The juice percentage is fairly high and the flesh tender but very variable in flavour, depending largely on the degree of pigmentation. The most intensely coloured have the strongest characteristic blood orange flavour. When harvested before January, the Moro is often very acidic. As peak maturity is reached the acid level falls and flavour improves. After February the fruit loses quality and does not ship as well. The processed juice has a colour medium burgundy in intensity that is much appreciated throughout Italy but barely known elsewhere.

SANGUINELLI

The Spanish Sanguinelli originated as a bud mutation of Doble Fina discovered in 1929 at Almenara in Castellón Province.

Sanguinelli trees are of medium size, vigorous, thornless and productive. The fruit, of small to medium size, is oval in shape but often lop-sided and with a persistent style.

The rind is extremely smooth, shiny and usually very well pigmented over much of its surface; it is moderately thick and leathery, but is not particularly easy to peel, and a significant amount of albedo is often retained on the segments. Although not as deeply pigmented internally as might be expected from external appearances, it has nevertheless good attractive internal 'blood' characteristics, although the strongest concentration of pigmentation is adjacent to the segment walls and much of the flesh is almost unpigmented.

The flesh is tender and has a high juice content; the flavour is sweet and less acidic than Doble Fina but has the same pronounced blood orange flavour. Most fruits have no more than one or two seeds but few are completely seedless.

SANGUINELLI
A pigmented orange of Spanish origin

Sanguinelli matures in January and remains in better condition on the tree than Doble Fina without losing quality; it also stores far better after packing.

Although it was widely planted in Spain (mainly at the expense of Doble Fina) its importance, like that of all the Sanguina varieties in Spain, has diminished considerably in the past two decades in favour of other finer quality varieties such as satsuma, Clementine and navel oranges.

SANGUINELLO

The common form, Sanguinello Comune, is an ancient variety of unknown origin and for a long period was the leading Sicilian blood orange variety. There are two derivatives: Sanguinello Moscata and Sanguinello Moscata Cuscuna.

Trees are moderately vigorous, rounded at the top and achieve medium to large size. All three varieties have medium sized fruit which is roundish in shape. The Sanguinello Moscata fruit has a furrowed and somewhat rough stem-end compared with the other two, while the Sanguinello Moscata Cuscuna has the thinnest and smoothest rind, the others being relatively thick.

SANGUINELLO
An important Italian pigmented orange variety

All Sanguinello selections have a variable degree of rind pigmentation, although none develop to the intensity of some other blood oranges such as Moro or Sanguinelli. One characteristic shared by the three Sanguinello selections is the oiliness developed during peeling, which is otherwise not at all difficult.

Sanguinello Moscata Cuscuna matures in January; Sanguinello Moscata and Sanguinello Comune are relatively late, reaching acceptable quality in February and March. Although of better quality than Moro, being sweeter, less acidic and more consistent, Sanguinello fruit never approaches the fine flavour of Sicily's best orange, Tarocco, and the group as a whole is losing ground to this superior variety.

TAROCCO

Originated from the oldest and now rarely propagated light blood Sicilian variety Sanguigno (not to be confused with Sanguinello), the Tarocco makes only a medium sized, somewhat irregular shaped tree which is only moderately productive.

Several selections have been propagated throughout southern Italy as follows.

Tarocco del Francofonte This is the commonly grown selection in Sicily. It has medium to large fruit of distinctive shape, being rounded with a slightly pointed collar terminating in a medium sized neck and having an overall appearance not unlike the Minneola tangelo. The rind achieves good orange colour when mature but has no red cast or blush, is finely pebbled, moderately thick, fairly loosely adhering and easily peeled with barely any release of rind oil. The segment walls are tender and the lightly pigmented flesh is soft and rag-free in texture.

The Tarocco is often not at its best until late January, but when fully mature the flavour is quite outstanding. Rich and fragrant with an ideal balance between sweetness and acidity, it is among the best of Mediterranean oranges. Furthermore, it is virtually seedless and stores well although losing quality if left too long on the tree.

Tarocco del Muso This selection is different from the common Tarocco in having a more pronounced neck.

Tarocco Rosso This strain has a noticeable red blush on the rind but is otherwise similar to the Tarocco del Francofonte.

TOMANGO

The origin is not known with certainty but it was first propagated on a commercial scale on H. L. Hall's Mattafin Estate, Eastern Transvaal, South Africa, in 1916.

The trees are slow growing and have an upright habit with leaves arranged in clusters, and they are slow to come into production. Of only limited adaptability, the Tomango has the serious shortcoming of producing predominantly small fruit, albeit of good quality.

Fruit is slightly oval in shape with a smooth rind texture, pale orange colour and devoid of any blush. The thin rind is fairly easily peeled and the flesh is particularly tender and juicy. Seeds are small and few in number, typically only one or two in each fruit.

TAROCCO
Italy's finest quality orange variety

TOMANGO

Although Tomango is a light blood variety, this trait is not normally expressed since production is predominantly in the warmer regions of South Africa. When grown under semi-tropical conditions it has excessive vegetative growth, poorer internal quality and a yellowish-orange rind colour. In cooler sub-tropical Mediterranean climates it matures very soon after the navel, with which it cannot compete in terms of fruit size and quality.

WASHINGTON SANGUINE OR DOUBLEFINE AMELIORÉE

(Grosse Sanguine)

Washington Sanguine originated as a bud mutation of Doble Fina found by B. Ferrer at Sagunto, Valencia Province, Spain. Like Doble Fina, Washington Sanguine trees are small but lacking in vigour, although they bear well in their early years and are productive.

Fruit size is much better than Doble Fina, being medium to large, round to oval in shape, often

WASHINGTON SANGUINE OR DOUBLEFINE AMELIORÉE

with a small concealed navel, and a slightly pebbled, medium thick rind. Often there is either a persistent style or a protruding 1 to 2 mm stylar scar.

Usually only very slightly blushed (with many showing no trace) the rind is not so leathery as Doble Fina or Sanguinelli but is only moderately tightly adhering and peels easily. The degree of flesh coloration depends upon climatic factors but a fruit with moderately pigmented flesh is rare – most barely have more than the slightest trace of 'blood' and many are completely 'white', even when produced in the cooler regions of Spain and Morocco.

The flavour is less acidic and sweeter than Doble Fina but has none of the distinctive blood orange taste which characterises other varieties such as Doble Fina, Sanguinelli and Moro. The fruit hangs better on the tree than Doble Fina but has poor juice processing quality.

ACIDLESS OR SUGAR ORANGES

There is a small but fairly widely grown group of orange varieties which have extremely low acid levels. They are greatly appreciated by many people, particularly young children in the countries in which they are produced, but only small quantities enter into world trade; as a result throughout much of Europe, North America and Australasia for example they are hardly known.

Acidless oranges are believed by some to have special therapeutic properties and are often eaten by the sick and infirm and by those who consider they are prophylactic. They are unsuitable for processing.

These are known under various names throughout the world:

Dolce	Italy
Douce, De Nice	France
Lima	Brazil
Lokkun, Tounsi	Turkey
Meski	North Africa and Middle East
Succari	Egypt
Sucreña	Spain

While there are slight differences between these varieties, they are essentially very similar, except for the time of maturity of some selections.

SUCCARI

Of unknown origin, this Egyptian variety is produced extensively, and is very popular locally and in neighbouring countries. The trees are vigorous and very productive.

Maturing just as soon as sugar levels have reached an acceptable level and the rind has developed sufficient colour, the Succari is harvested from October onwards and hangs well on the tree without becoming puffy or granulated. Small to medium in size, the fruit is round in shape but has no distinguishing features. It is smooth, usually well coloured and has a thin rind which is not easily peeled since it is tightly adhering. The rind may become creased if the fruit is left on the tree too long. Succari is very seedy, often exceeding 25 seeds per fruit.

The flesh is of good orange colour, very juicy, tender but extremely insipid and of unacceptable quality to those not acquainted with its unfamiliar flavour. The sugar levels in the juice of around 10 per cent are not unlike those of 'sweet' oranges but, whereas the latter might be expected to have an acid level of about 1 per cent, the Succari has only 0.1 per cent. It is not simply the ratio of sugar to acid that is different; it is a lack of the familiar 'orange' flavour which characterises this and other acidless varieties.

Annual production of Succari in Egypt is estimated at 4 million 15 kg cartons.

SUCCARI
An Egyptian acidless orange variety

SUCREÑA

In Spain the acidless orange variety Sucreña (Real or Imperial Grano de Oro) is less popular locally than it was in the past and production has virtually ceased. The same applies in Sicily with the Vaniglia variety.

LIMA

There are several selections of the Lima variety in Brazil which can be divided into two groups.

Lima and Piralima Both these selections mature very early, and are harvested from March onwards in Sao Paulo State. The Piralima, selected by P. W. Cabral de Vasconcellos in the Piracicaba area, is flatter, smoother and paler in colour than the Lima, with a lower seed content and better productivity.

Lima Verde Lima Verde is much later maturing and still green when the Lima is fully coloured (hence its name). It eventually develops pale orange rind colour and is of otherwise similar quality and appearance.

Lima oranges in Sao Paulo State may form only 6 per cent of the total orange production, but this represents over 0.5 million tons (or 35 million 15 kg cartons) and reflects the enormous popularity of acidless oranges in Brazil.

MOSAMBI

The Mosambi variety, probably of East African/Portuguese origin and popular throughout Pakistan and India, is often mistakenly regarded as an acidless type. It has some acidity when grown elsewhere, but in the Indian subcontinent the climate results in a virtually acidless fruit being produced.

AN ORCHARD OF CLEMENTINES, CAPE PROVINCE, SOUTH AFRICA

THE MANDARIN

Like that of other citrus species the precise origin of the mandarin is far from certain but is believed to be either north-east India or south-west China. It is known by various names as follows:

Mandarin	English
Mandarino	Italian and Spanish
Chu, Ju, or Chieh	Chinese
Mikan	Japanese
Santara	Indian

The mandarin has probably been cultivated in China for several thousand years, and the earliest reference to this fruit dates back to the 12th century BC. From its region of origin, the mandarin spread throughout much of south-east Asia, and to other parts of India. By the tenth century AD the mandarin was widely cultivated in the southern prefectures of Japan.

However, the origin of the Unshiu mikan or satsuma can be traced to the early 16th century from the Tsao Chieh mandarin imported from Wenzhow, Zhejiang Province, China, probably 1,000 years earlier. (Unshiu is a Japanese corruption of Wenzhow.)

It was a further 300 years before mandarin distribution on a global scale began when two mandarin varieties from Guangzhow (Canton) were imported into England by Sir Abraham Hume in 1805. From this introduction trees were sent to Malta, then to nearby Italy; it is believed the Mediterranean mandarin evolved under cultivation in Italy shortly thereafter.

Mandarin trees and their hybrids are usually the most cold-resistant of all commercially grown citrus although there are some, such as the Temple tangor, which are less hardy than oranges. However, the fruits of mandarin suffer more frost damage than most oranges and grapefruit.

The mandarin has a wide range of adaptability and is grown under desert, semi-tropical and subtropical Mediterranean climatic conditions. Despite this, different varieties of mandarin are very specific in their climatic requirements for good production and quality. For example, the Ponkan, Tankan, Ellendale and Dancy are best suited to semi-tropical conditions and are rarely grown alongside the satsuma (Unshiu) mandarin which is at its most productive and achieves its finest quality only in regions which have cold winters. The most demanding of mandarins in its climatic requirements is probably the Clementine which still has a very restricted distribution, limited almost entirely to the coastal areas of Morocco, Spain and Corsica. Many mandarins – in particular the Mediterranean mandarin and Dancy – and their hybrids tend to alternate-bearing, when large crops of very small fruit are followed by light crops of large poorer quality fruit. Often other varieties are recommended as pollinators to encourage better fruit set while several horticultural practices such as girdling of the stems, spraying with growth regulators and hand-thinning are employed to affect the irregular bearing behaviour.

Features shared by almost all mandarins are their relatively short harvesting season and their susceptibility to injury during picking, packing and transporting to market. The rind is more brittle and sensitive to injury with a tendency to become 'puffy', while internally the flesh loses acidity and juice content and becomes insipid if left on the tree for only a relatively short period after reaching peak maturity. However, if handled with care and harvested before becoming overmature mandarins may be successfully stored for many weeks and sometimes longer.

Referred to by a variety of names, probably conjured up by advertising executives and ranging from 'soft-citrus' and 'kid-glove' to 'zipper skinned' and 'easypeeler', mandarins have long been appreciated for their fine distinctive and sweet flavour. Yet it is their ease of peeling which is the most common and outstanding feature and has led to their popularity in many countries throughout the world.

The name 'tangerine' has often been synonymous with mandarin, particularly in the United States where it was first used in conjunction with the Dancy variety. Later it was extended to other mandarins with similar deep reddish-orange rind. However, there are conspicuous inconsistencies: some varieties with colour no deeper than oranges are referred to as tangerines while the variety Temple with even redder rind colour is more often than not called Temple orange! To complicate matters further it was until quite recently the practice on European wholesale markets to refer to all seedy 'easy peelers' as mandarins. Fortunately, that is no longer the case and perhaps the time has come to refer to all loose-skinned citrus fruit as mandarins prefixed by their specific name, e.g. Clementine mandarin and satsuma mandarin.

Although mandarins are widely distributed, annual world production of mandarins is estimated at only around 7 million tons, which is far less than that of oranges at 44.5 million tons. However, it should be remembered that as much as 15.8 million tons of oranges are used for processing into juice.

There have been several attempts to group mandarins into different categories or species; the one followed here is that of R. W. Hodgson:

Citrus unshiu	Satsuma mandarin or Unshiu mikan
Citrus deliciosa	Mediterranean mandarin
Citrus nobilis	King mandarin
Citrus reticulata	Common mandarin

The first three are well-defined mandarin groups with small but nevertheless distinct varieties within each but the fourth is an extremely wide collection of varieties, many of them natural or man-made hybrids.

SATSUMA MANDARIN

(Citrus unshiu)

The satsuma mandarin or Unshiu mikan almost certainly originated in Japan as a nucellar seedling from the Tsao Chieh mandarin imported from Wenzhow, China, probably in the mid-sixth century AD. It was given the name satsuma in 1878 by the wife of the United States Minister to Japan, General Van Valkenberg, Satsuma being the former name of the prefecture now known as Kagoshima on Kyushu Island in western Japan.

Satsumas are grown principally in Japan and Spain and to a far lesser extent in other countries worldwide. There are satsuma industries on Cheju Island, Korea, and in many Mediterranean coastal areas of Turkey, particularly near Izmir, and along the Black Sea coast near Rize. Near Batumi in the USSR satsumas are grown on a fairly extensive scale. A very small area is devoted to them in the Central Valley, California, and in coastal regions of Argentina, Uruguay and South Africa. The satsuma has been introduced to China and now predominates in the eastern province of Zhejiang and in the lower and middle Yangtze Valley. However, the Ponkan and Tankan are still China's most extensively grown mandarins.

Of all mandarins, none has more exceptional cold-hardiness than the satsuma and this is particularly so when grown on trifoliate orange rootstock.

The climates in Spain and Japan are characterised by cold winters – particularly in many parts of the citrus-producing areas in Japan. However, the summer months in the two countries are quite different. Japanese summers are humid with high rainfall during the early months, while in Spain along the Levant coast there is virtually no rainfall during the period of fruit growth and the humidity is fairly low. Despite these marked contrasts, both countries produce satsumas of excellent quality.

Both industries were founded on the same variety, Owari. Spain has not developed more than one other improved strain of satsuma – although a few others have been recognised – having concentrated its efforts on finding new Clementine

SATSUMA MANDARIN OR UNSHIU MIKAN

varieties. In Japan, on the other hand, many different selections have been propagated, which has extended the Unshiu season considerably with the development of both earlier and later maturing varieties. The season has been further extended with the cultivation of Unshiu mandarins in heated plastic houses, which advances the season by as much as four months or more. Sophisticated post-harvest storage programmes have been developed for citrus generally and these have enabled Unshiu fruit to be stored for several months. The net result of these developments now permits fresh Unshiu to be available on the Japanese market for as long as eight months and for mandarins of one variety or another for virtually 12 months of the year.

It is perhaps appropriate here to sound a note of caution. The unauthorised movement of citrus budwood material from one country to another (and sometimes inter-regionally within a country) always carries with it enormous risks of transferring pests and diseases. This is especially so with the satsuma from Japan where most of the trees are infected with citrus canker A (Cancrosis A) to which the variety is largely tolerant.

Satsuma mandarins are well suited to processing for juice but, being seedless, the segments are especially appropriate for canning in syrup or natural juice; both Japan and Spain have built a substantial export trade in canned satsuma segments.

Satsuma production in Spain and especially in Japan has been declining in the past decade. In the mid-1970s around 170,000 ha were planted to Unshiu mandarins in Japan, producing over 3.6 million tons, but this has declined to approximately 120,000 ha and around 2.2 million tons at present and is forecast to decline further in the near future.

Satsuma varieties can be conveniently divided into two main groups on the basis of their distribution in Spain and Japan with the variety Owari common to both.

OWARI

Present-day Aichi Prefecture in Japan was formerly known as Owari, whence this variety probably originated. It became the principal Unshiu variety in Japan during the early part of the 20th century and was imported into Spain around 1925.

Owari trees are vigorous, very productive and come into bearing at an early age. Although fairly early maturing the Owari is preceded by several weeks in Japan by the 'Wase' (meaning early) groups of Unshiu varieties and in Spain by the variety Clausellina. Fruit size is medium to large for a mandarin and is always completely seedless. The Owari fruit is of larger size and much flatter shape and Japanese-grown fruit usually has a smoother peel texture than Spanish-grown because of the more humid growing conditions. Flavour is also affected by climatic conditions; Japanese Owari is sweeter, less acidic, and has the more distinct and delicate flavour. In both countries the fruit cannot be held on the tree too long after reaching maturity without becoming puffy and rapidly losing quality.

OWARI
The fruit is internally mature before the rind is fully coloured

At present, Owari is planted on around 17,000 ha in Spain and 13,000 ha in Japan. This represents about 90 per cent of Spain's 400,000 ton satsuma crop and 15 per cent of Japan's 2.2 million ton Unshiu mikan production respectively.

Other Spanish Satsuma Varieties

CLAUSELLINA

A bud mutation from Owari, this variety was discovered as recently as 1962 in Almazora near Castellón. It is the earliest of Spain's two commercial satsuma varieties, reaching acceptable maturity about three weeks ahead of the Owari; however, the quality when fully mature is lower, being less sweet, but having otherwise similar characteristics. Although satsumas generally are not being planted in Spain at the same rate as Clementine varieties, Clausellina is much more popular than Owari on account of its earlier maturity.

The trees tend to grow more slowly than Owari and their final size is noticeably smaller, which facilitates their being planted more closely so production is similar to the Owari on an area basis.

PLANELLINA

R. Bono, L. Fernandez de Cordova and J. Soler of IVIA, Moncada, Valencia, have quite recently reported the Planellina, a bud mutation of Owari, as being a very early maturing satsuma – five weeks ahead of Owari and two weeks ahead of Clausellina – and having similar internal quality to Owari. It is juicier, has thinner rind, and is slightly larger in size than the Clausellina.

Other Japanese Unshiu Varieties

Japanese Unshiu varieties are classified into five groups according to their time of maturity. The leading varieties are as follows.

Very Early	*Early*	*Midseason*
Miyamoto	Miho	Nankan No 4
Oura	Miyagawa	
Tokumori	Okitsu	
Ueno		
Yamakowa		

Common	*Late*
Hayashi	Aoshima
Sugiyama	Juman
Ohtsu No 4	Imamura
Owari	

Very Early Group

Recent reports indicate the discovery of exceptionally earlier maturing Japanese Unshiu selections which may be harvested in September compared with early October for 'Wase' varieties. All have dwarf habit and include the varieties Miyamoto, Oura, Tokumori, Ueno and Yamakowa.

Early Group

MIHO

(Miho Wase)

This nucellar seedling was derived at the same time as Okitsu in 1940 but has never been commercially developed to any extent, although it has similar fruit characteristics, with the exception of slower colour break.

The tree is more vigorous than Okitsu. It has performed well in recent trials in Cape Province, South Africa, producing fruit much earlier than Owari and of superior quality.

MIYAGAWA

(Miyagawa Wase)

This is the most widely grown Unshiu variety in Japan, occupying about 25,000 ha. Originating as a bud mutation on a tree of the seedy Zairai variety in Fukuoka Prefecture, it was introduced in 1923 by T. Tanaka. Miyagawa fruit is larger than Owari, the rind is somewhat smoother and thinner, and it matures three weeks earlier.

The flavour is quite sharp because of a high acid level, and sugar levels are not particularly high.

OKITSU

(Okitsu Wase)

The second most popular satsuma variety in Japan and grown on about 20,000 ha, Okitsu was produced as a nucellar seedling by controlled pollination from a Miyagawa fruit in 1940, but was not developed on any scale until the 1960s. It has recently been planted on a limited area in Spain.

It is of similar size to the Miyagawa, but slightly flatter, and its flavour is better because the acidity is lower and sugars slightly higher. It matures seven days earlier than Miyagawa and has better storage characteristics.

Midseason Group

NANKAN NO 4

Nankan No 4 originated as a bud mutation on an Owari tree on S. Yakushiji property in 1925. It is not grown on as extensive a scale as other varieties but is particularly popular in its native Ehime Prefecture.

The tree has good vigour and productivity. The fruit is large for Unshiu, and is particularly sweet with adequate acidity to give a balanced flavour.

Common Group

HAYASHI

Grown on approximately the same scale as Owari in Japan, this variety originated on the property of B. Hayashi in Wakayama Prefecture in 1920. It is very vigorous and more upright in growth habit than most other varieties. The Hayashi has a high sugar content, is moderately acidic, and is regarded by many as the finest flavoured of all Japanese Unshiu varieties. It stores exceptionally well.

SUGIYAMA

At one time the leading Japanese satsuma variety, it originated as a bud mutation of Owari on J. Sugiyama's property, Shizouka Prefecture, in 1934.

The trees are moderately vigorous but very productive, and the fruit matures in November.

The fruit is the same size as Owari and very attractive with a particularly smooth rind. On account of low acidity and only moderate sugar content, the fruit lacks flavour and cannot be stored for any length of time.

Late Group

AOSHIMA

Discovered in 1950 as a bud mutation on an Owari tree, but as yet planted on only a small area, this exceptionally late maturing Japanese variety has an attractive smooth rind which is more tightly adhering than other varieties, but is not difficult to peel. It has a high acid level, but this is coupled with an exceptionally high sugar content.

This variety is grown mainly in Shizouka Prefecture with the object of long-term storage. Many fruits have a small but noticeable navel.

JUMAN

The Juman originated as a bud mutation of Owari on the property of K. Juman in Kochi Prefecture on Shikoku Island in 1953. It has always been regarded as having potential and is presently being planted on an increasingly large scale. Juman has exceptionally high sugar levels of 12 to 13 per cent and is well suited to prolonged storage. It has given rise to a bud mutation, Ohtsu No 4, which is of similar high quality but slightly earlier maturing.

MEDITERRANEAN MANDARIN

(Citrus deliciosa)

This was the first mandarin introduced into the Mediterranean basin from China in 1805 via England, to Malta and finally to Italy, before being distributed throughout the region and later worldwide. The Canton mandarin (Szekan or September mandarin) grown extensively in Guangdong Province has many characteristics in common with the Mediterranean mandarin and it is believed by many authorities that it was from this variety that the first importation was made.

It is known by many different names, some of which include:

Mediterranean	England
Ba Ahmed	Morocco
Bodrum	Turkey
Avana, Palermo and Paterno	Italy
Baladi, Yusuf Efendi	Egypt and Middle East
Comun, Valencia	Spain
Setubal, Gallego	Portugal
Thorny	Australia
Mexirica do Rio	Brazil
Chino, Amarillo	Mexico
Willowleaf	USA

The tree is slow growing and of medium size with drooping branches, is nearly thornless, and has very small narrow leaves. It is fairly cold-resistant and strongly inclined to alternate-bearing.

The fruit is small to medium in size and fairly oblate in shape, with a small but heavily furrowed neck, and often with a small, sometimes concealed navel.

The rind, which is yellowish-orange when fully mature, is thin, smooth, often creased and very loosely adhering. It peels with the utmost ease, and the rind oil has a distinct fragrant aroma unique to Mediterranean mandarin. The segment walls are moderately tough, but the orange-coloured pulp is tender, melting and very juicy, with a distinct mild, aromatic and sweet flavour. A midseason

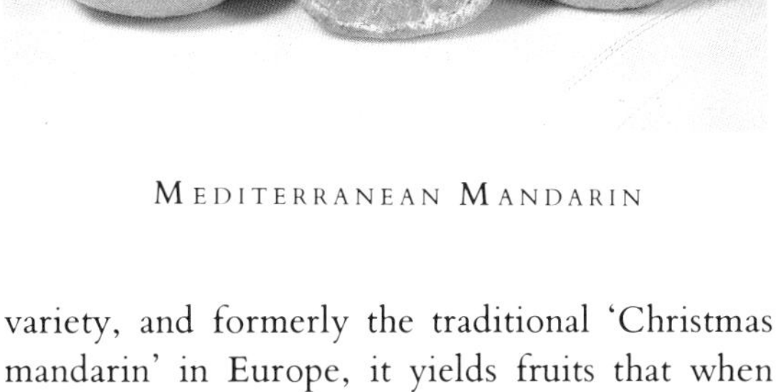

MEDITERRANEAN MANDARIN

variety, and formerly the traditional 'Christmas mandarin' in Europe, it yields fruits that when overmature quickly become puffy (with the rind barely adhering to the segments) and lose juice and acid content. Unlike the satsuma mandarin, for example, it does not ship or store well.

In Sicily there are three improved selections:

Avana di Palermo It is characterised by its smooth peel, pale orange-coloured rind and small segments.

Avana di Paterno This strain has a distinct neck, rough peel texture, deep orange-coloured rind and large segments.

Tardivo di Ciaculli This has been most popular in recent years. It is flatter, sweeter flavoured and as much as two months later maturing.

During the last two or three decades production of the Mediterranean mandarin has declined very significantly in favour first of the satsuma and, latterly, the seedless Clementine.

There are two important by-products of the Mediterranean mandarin: the rind oil which is used in the manufacture of perfumes and to flavour soft drinks; and oil of petitgrain which is distilled from the leaves and twigs, and is also used in perfume production.

KING MANDARIN

(Citrus nobilis)

This small and unimportant group probably originated from a natural mandarin–orange cross, and comprises the two varieties: King, from Cambodia and Vietnam; and Kunenbo, from Japan. They differ in size and peel texture, Kunenbo being smaller and having a smoother rind texture as well as having a somewhat bitter flavour.

KING

(King of Siam)

An old variety of unknown origin, King was first sent to California in 1880 from Saigon, Vietnam, and thence to Florida.

The tree is fairly vigorous but is of only medium size when mature, having a few thick branches. However, in California King has sparse foliage and tends to die back. It is moderately thorny, has large dark green narrow leaves and is very productive. The wood is brittle with a tendency to break under the load of a heavy crop. King shows a tendency to alternate-bearing. It is fairly cold-resistant, but more susceptible to damage than other mandarins.

Fruit appearance is distinctive, being large in size like many oranges, roundish in shape with a slightly flattened apex and base, and with a thick to very thick rind which is exceedingly rough and uneven in texture. Rind colour is yellowish-orange to orange at maturity and is very easily peeled but excessively oily. King is one of the most susceptible of all citrus fruit to any handling damage.

The segments give the appearance of being dry, as is often found with Pixie and the Mediterranean mandarin. The flesh has deep orange colour, and is low in juice, but tender. The segment walls are tough and the fruit moderately seedy, averaging about 12 seeds per fruit.

Although the juice has moderate acidity when mature it also has high solids so the result is very sweet but lacking in any distinction and it cannot be compared with better quality varieties, such as the Hernandina and Nour Clementines and Fortune.

King is extremely late maturing and is not considered at its peak until spring the following year. Grown extensively in south-east Asia, it has never been of any commercial significance elsewhere. However, as a parent of the hybrids Encore, Kinnow, Wilking and Kara, it has found a place in citriculture history. To date none of its progeny have achieved lasting importance, although all at one time held some promise.

Some observers have recognised the similarities between King and the Argentinian variety Malaquina (or Campeona) but in my opinion they are probably unrelated.

COMMON MANDARIN

(Citrus reticulata)

Since this group is so extensive and has many different varieties, it is difficult to list characteristics which are common to all, except to note that they tend to have rind which is somewhat more tightly adhering than that of the satsuma and Mediterranean groups. Also they are less likely to become excessively puffy late in the season. This group contains some of the most important mandarins grown worldwide such as the Clementine, Dancy, Murcott, Ponkan and Tankan as well as other mandarin hybrids like Ellendale, Nova, Fortune and Robinson.

CLEMENTINE

Clementine

There are two suggested explanations of the origin of the Clementine. According to one, it resulted from an accidental cross between the common Mediterranean mandarin with pollen from an ornamental sour orange known as Granito, first noticed and named by Father Clement Rodier (hence the name Clementine) who discovered it in the garden of an orphanage at Misserghin near Oran, Algeria, around the end of the 1890s. However, little credence is now given to this possible origin. Instead some authorities believe it is virtually identical to the variety known as the Canton mandarin widely grown in Gwangxi and Guangdong Provinces of China.

In the Mediterranean region, particularly in Morocco and Spain, the Clementine has become the most popular and fastest expanding mandarin variety during the past three decades. Only true seedless Clementines have been developed in Spain and by diligent observations many new exciting mutations have been discovered recently. This has extended the Clementine season from its former two-month period from November to December to cover the period from mid-October to mid-February. The late maturing Clementine hybrid Fortune, of similar quality to the Clementine parent, has resulted in a further extension until mid-April in Spain.

Spain's annual Clementine crop of around 650,000 tons now accounts for about 20 per cent of the country's combined orange and mandarin production of 3.2 million tons and is forecast to rise in the foreseeable future. Clementine production in Morocco is around 400,000 tons, of which about half is exported mostly to western Europe.

The important Clementine varieties currently planted and those showing promise are as follows.

ARRUFATINA

The Arrufatina is of recent origin, having been discovered as a bud mutation of Nules at Villarreal, Castellón Province, Spain, in 1968.

The trees are thorny when young, have good vigour and develop into large specimens. They are also productive and quickly come into bearing. Fruit quality is good without matching that of Fina or Nules for sweetness and flavour, but it is less acid than Oroval which it is now replacing. Fruit size is comparable with Oroval.

It matures in early November, two to three weeks earlier than Nules, but has to be harvested at colour break and degreened since natural colour development is slow and the rind tends to become puffy in the intervening period.

BEKRIA

Bekria (meaning 'early') refers to a group of very low acid Clementine selections produced on a very limited scale in Morocco. When they are harvested in September the rind is still green having barely reached colour break, the acidity is already low but the sugar level is no more than moderate, resulting in an insipid flavour. Moreover, the flesh, although reasonably juicy, seed-free and of fairly tender texture, is pale orange. Because it is so early, it commands good prices on the local market in Morocco and in France where it is much appreciated by expatriate North Africans.

BEKRIA
The earliest maturing Clementine, which is harvested while still green

ESBAL

A bud mutation discovered on a Fina tree in 1966 near Sagunto in Valencia Province, Spain, Esbal has some outstanding features. The trees are vigorous and attain very good size, and they quickly come into production.

Although fruit size is only slightly larger than Fina, Esbal has the same excellent flavour and extremely tender flesh. It is of outstanding colour, and may be harvested a week or so before the Fina, but does not hold well on the tree after attaining full maturity as the rind is susceptible to rain and dew.

Unlike other seedless Clementine varieties, it has the important advantage of setting a good crop without needing hormonal or girdling treatments.

FINA

(Algerian)

First introduced into Spain in 1925, probably from Algeria, the Fina laid the foundations on which the country's Clementine industry developed. Until the early 1960s only Fina Clementine was grown on any scale in Spain. All other Spanish Clementines are derived from the Fina either directly or via one generation.

Fina trees are vigorous, dense and large and have good productivity. Although relatively later maturing by as much as four weeks compared with the early selections such as Marisol and Oroval, it is still the finest quality Spanish Clementine and is the one against which others are compared. Unfortunately the fruit is very small, much of the crop being below 60 mm in diameter (averaging 50 mm), with the result that market returns on a high proportion of smaller fruit cannot compete with other selections which produce larger if somewhat inferior fruit.

The rind is particularly smooth, and the fruit has excellent organoleptic characteristics: high juice content, very tender and sweet with good acid level but high sugar to acid ratio. It has the strong, pleasant aroma which typifies the Clementine.

Fruit may be left on the tree for a relatively long period without noticeable quality deterioration. It is recommended for planting only in areas where climatic conditions favour the production of comparatively large size fruit. The Clementine selection in California known as 'Algerian' is most probably the same as Fina in Spain. The Fina is no longer planted in Spain because of fruit size problems but around 10,000 ha are in production. Along with Nules, it is still the most extensively grown Clementine variety in Spain.

GUILLERMINA

This Fina mutation, discovered at La Llosa in Castellón Province, is still in the early stages of evaluation, but the trees have good vigour and productivity.

External colour is outstanding, being reddish-orange while internally the segments are deep orange, tender and very juicy, with excellent flavour and aroma. However, fruit size is only a little larger than Fina.

HERNANDINA

Discovered in 1966 as a bud mutation of Fina at Picasent in Valencia Province, the Hernandina is an exciting selection at present being extensively planted in the late areas of Spain.

Tree characteristics are almost the same as Fina, and so too are those of the fruit, with one important exception: the Hernandina's external colour develops two months later than the Fina. It is not harvested until mid-January and can be held in good condition and without quality deterioration until late February or early March.

Colour development is characteristically incomplete on a significant percentage of fruit, with a small but acceptable area of the rind at the stylar-end remaining slightly green. Somewhat surprisingly the internal maturity is reached not more than one or two weeks later than the Fina and remains outstanding for an additional three months.

The Hernandina does not store well after harvest and may develop granulation if held on the tree past peak maturity. Nevertheless, price realisations on European markets have been most rewarding and have encouraged current planting rates of over 100,000 trees per year.

MARISOL

This most promising of Clementine selections originated as a bud mutation on Oroval in 1970 at Bechi in Castellón Province. Tree and fruit characteristics are indistinguishable from Oroval, with one significant exception: Marisol matures at least two weeks earlier than Oroval and is therefore as early as the Owari satsuma and seems destined to make inroads into these two varieties. This is already evident from its current popularity, with plantings of around 250,000 trees per year throughout Spain (or 15 per cent of all mandarins).

MONREAL

Discovered in 1940 in the orchard of Vincent Monreal at Perregaux, Oran, Algeria, this variety has now been almost entirely superseded by improved seedless Clementine selections.

Being self-compatible and without cross-pollination this fruit is regularly seedy, with as many as 20 seeds per fruit. As is often the case with other citrus varieties, seedy fruit tends to be both larger and slightly sweeter than seedless selections.

However, the introduction of cultural techniques to enhance fruit set of seedless Clementines has resulted in the once reliable bearing characteristic of the seedy Monreal no longer being regarded as essential to good production.

MONREAL

Nour

There is little information available regarding the origin and tree characteristics of this increasingly important Moroccan Clementine selection.

It is later flowering than Fina by about 10 to 15 days and its maturity is significantly delayed. Whereas Fina may be harvested in late November onwards in Morocco, Nour is not mature until mid-January and may be picked without loss of quality until late February or early March – about the same period as the Spanish Hernandina.

Nour fruit has good colour, better than that of Hernandina, but the rind texture is somewhat coarse. The flesh is tender, juicy and of very good flavour. Production in recent years has risen to around 2,500 tons in 1989 and is forecast to double in three to five years.

Nules

(Clemenules, De Nules)

The most popular Clementine selection in Spain where it constitutes around half of current plantings, Nules was discovered near the town of the same name in Castellón Province as a bud mutation on a Fina tree in 1953.

Like the Fina, Nules trees are vigorous, attain large size, and are very productive, outyielding the Fina by about 10 per cent. Moreover, the fruit is significantly larger (although somewhat smaller than the Oroval), maturing only a few days later than Fina, in late November. An important characteristic of Nules is the extended period over which the fruit retains its quality, making it possible for fruit to be harvested until the end of January, if climatic conditions are favourable.

The extended harvesting period is made possible by up to three fruit sets, the fruit becoming more coarse and larger with each set. Picking selectively is therefore an essential part of good management of Nules orchards. Packers and shippers will commonly pay a 15 to 20 per cent premium for Nules over Oroval, so much better is the quality.

Oroval

Oroval, a bud mutation of Fina, was found in 1950 at Quart de les Valls in Valencia Province, Spain. The trees are vigorous, well developed but thorny, although this characteristic declines with age. The fruit is only slightly larger than Nules and matures fully three weeks earlier. However, it has two important disadvantages from a production point of view: poor hanging ability because the rind, which has a somewhat more pebbly texture than Nules, becomes excessively puffy with delayed harvest; secondly, a rind which is susceptible to what is known locally as 'water spot' following heavy rains, which causes the fruit to drop to the ground.

Although the flesh is reasonably tender and even more juicy than Nules, it is more acidic despite having good sugar levels. The urgency with which producers harvest the Oroval is sometimes reflected in poorer than optimum quality. This and other shortcomings have been noted by producers and are reflected in current plantings: only 1 per cent of all Clementines are of this variety. However, there are an estimated 7,000 ha in production at the present time.

SRA Selections

In the 1960s, as part of the programme to develop a viable citrus industry on the island of Corsica, France, the Citrus Research Station at San Guiliano (Station de Recherches Agrumicoles – SRA) selected different strains of Moroccan Clementines which were prefixed 'SRA'. One of them, SRA.63, was used as the basis for the successful if small Clementine industry which exists today on Corsica.

Unfortunately, despite some outstanding characteristics all the SRA strains had the serious defect of producing unacceptably small fruit and have not found a place in countries outside Corsica save in South Africa where the SRA.63 has been developed on a small scale. Most SRA selections are of excellent quality, on a par with the Fina from Spain.

Other Mandarins

This is a diverse sub-group, many of which are of hybrid origin, but all share the characteristic of being relatively easily peeled.

CRAVO

(Laranja Cravo)

Probably of Portuguese origin and similar in many respects to the other Portuguese variety Carvalhais, Cravo (and Honey tangerine) is the second most widely grown mandarin in Brazil, after Ponkan.

The trees are large and very productive, with mandarin-like leaves, but have a marked tendency to alternate-bearing.

The fruit is medium-large and flattened in shape with a somewhat rough, thickish, loosely adhering rind. Very early maturing, the flesh is tender and juicy, and the fruit stores well on the tree without becoming puffy. The flavour, however, is insipid but much appreciated on the Brazilian local market, and it is moderately to excessively seedy with never less than about ten seeds per fruit.

DANCY OR DANCY TANGERINE

(Obeni-mikan)

An old variety from Florida, it originated from a seed of the variety known as Moragne 'tangierine' in 1867 in a grove of Colonel F. L. Dancy at Orange Mills.

Dancy trees are vigorous and large (for a mandarin), thornless and with an upright growth habit. Although the tree has some cold-resistance, the fruit is susceptible to frost. It has a strong tendency to alternate-bearing and often produces heavy crops of small fruit. A midseason variety, harvested during December and January, it is the traditional 'Christmas tangerine' in the USA.

The overall shape of the Dancy is slightly flattened at the stylar-end but with a slight depression close to the stylar scar where sometimes a small navel is formed. The stem-end is sometimes slightly pointed and most fruits have a small neck, while the rind is smooth but often uneven when fully mature.

DANCY
An old variety from Florida

The rind is deep reddish-orange, thin, leathery and very easily removed, becoming extremely puffy shortly after full maturity. The segment walls are moderately tough but the flesh is tender, although not abundantly juicy. However, the juice is rich and sweet and yet has a high enough acid level to give the fruit a well-balanced sprightly flavour. It quickly loses quality if picking is delayed. There are never less than six seeds per fruit and the average sometimes exceeds 15.

The peel is very susceptible to tearing (or 'plugging') if not picked with care, and Dancy tangerines have a reputation for arriving in the market with high levels of decay.

Although it has been grown on a very extensive scale in Florida where it is particularly well adapted to the climate, production has declined dramatically in recent years from around 80,000 to about 10,000 tons. Only in the desert areas in California and Arizona is fruit of acceptable size and quality produced but sunburn of exposed fruit is a problem.

Dancy tangerines, produced in Israel, were found to be too seedy for the European market, and exports were curtailed soon after they began.

ELLENDALE

This important and high quality variety, discovered as a seedling by E. A. Burridge as early as 1878 at his Ellendale property, Burrum, near Bundaberg, Queensland, Australia, is obviously a tangor (orange–mandarin hybrid), judging by its size and its many mandarin characteristics, although its parentage is unknown. Several improved selections have been made and include Leng, Koster and Barlow.

Ellendales make larger sized, round-topped trees, thornless but prone to crotch splitting, although new selections are less affected.

The fruit is medium to large, sometimes very large, usually flattened in shape and is regarded as being a midseason maturing variety in Queensland. Slow to colour up, the rind eventually develops a good deep orange colour. Rind texture is smooth on all except extra-large fruit and, although thin – sometimes extremely thin – is easily removed without releasing much rind oil. It does not develop puffiness when left to hang on the tree.

Ellendale rind has a distinct tendency to split at the stylar-end, especially in hot humid growing conditions, causing the fruit to drop from the tree. This may occur over a long period leading up to harvest.

Of outstanding colour, the pulp is tender and extremely juicy and has a good, sweet, rich flavour. Acid levels are inherently high, so it is important that they are given sufficient time to fall to acceptable levels before harvesting. Selective picking is sometimes necessary to permit small fruit to achieve a good and acceptable sugar to acid ratio.

Seediness in Ellendales is variable and probably related to climatic conditions at flowering time. When cross-pollinated, fruits may contain as many as 20 seeds, but solid-block plantings often bear fruit which is virtually seedless.

Ellendale may be picked over a long period, which may be extended further with the use of gibberellic acid sprays applied shortly after colour break. The fruit also stores and ships well.

Ellendales produce their best quality on Troyer and Carrizo citrange, and trifoliate orange rootstocks, but in cooler areas this leads to high acid levels. Moreover, creasing becomes problematical late in the season on trifoliate orange. Sweet orange is also a successful rootstock for Ellendale in Queensland. There is an incompatibility problem with rough lemon which only becomes evident about ten years after planting, but trees are reported to grow well on Rangpur except following a replant. On account of the high acid levels in the fruit, Ellendale trees should receive less potassium than other mandarins and oranges.

For many years production was restricted to the Central Burnett District north-west of Brisbane as yield and quality problems occurred on coastal and more southerly areas of Australia. In the past decade production in Argentina and Uruguay has increased significantly and an export trade to Europe of good quality Ellendales has developed.

ELLENDALE
An important Australian mandarin

ENCORE

FAIRCHILD

ENCORE

A hybrid originating from a cross of King mandarin and Willowleaf mandarin by H. B. Frost in California, and released in 1965. The trees are medium size, the branches thorny and there is a strong tendency to alternate-bearing.

Encore has the shape and rind texture of the Dancy but without the short small neck. It is quite firm for a mandarin, and the rind is predominantly yellowish-orange in colour but blotchy with darker orange patches, and is very thin and tightly adhering. However, it is easily peeled with very little albedo retained on the segments.

When the segments are separated the fruit 'sounds' dry as is sometimes the case with an overmature Dancy or Mediterranean mandarin, and it has the same distinctive rind oil aroma of the latter variety. Segment walls are decidedly tough but the pulp is tender and juicy with a pleasant, sweet, well-balanced flavour. However, it is excessively seedy, averaging 25 or more per fruit.

Like the Fortune it reaches full maturity extremely late – later even than King – but has yet to be grown on any significant commercial scale save in Japan, where about 1,500 tons are produced in plastic houses, and harvested in March.

FAIRCHILD

The Fairchild is a Clementine–Orlando tangelo hybrid made by J. R. Furr at Indio, California, and released in 1964. It has good vigour and develops into a thornless, spreading and densely foliaged tree with good productivity.

The fruit is very early maturing, attractive and has many of the Clementine parent's characteristics of shape, rind texture and colour, although it is of slightly larger size. The fruit condition is usually firm. The thin, fairly tightly adhering rind is not readily removed, and breaks into small pieces. It is very oily to peel which makes contamination of the segments unavoidable.

The flesh may be rather coarse but is juicy and tender. However, there is a lack of flavour compared to Clementine or Nova. On some occasions an unusual and strange, somewhat tomato-like aroma is detectable.

Yields are increased with pollinators (e.g. Orlando tangelo) which results in the fruit being excessively seedy. Fairchild has yet to be produced anywhere on any significant scale, although it is assuming some importance in desert areas of California and Arizona. There, in years of heavy cropping, small fruit size is a problem.

FORTUNE

(Fortuna)

A hybrid of the Clementine (Algeria) and Dancy tangerine made by J. R. Furr in California in 1964, this highly promising mandarin is presently receiving much attention in Spain where it has performed well in the past decade. Elsewhere the Fortune has proved to be very unpredictable in terms of quality under a range of different climatic conditions.

The trees are vigorous, of medium size and very productive, bearing most fruits inside the canopy.

The fruit is of good size for a mandarin and in shape and peel texture more closely resembles the satsuma than Clementine, but is not nearly so soft as the typical satsuma. Colour is good and the rind, which is very thin and fairly tightly adhering, is easily and cleanly removed.

However, due to its extremely late maturing characteristic – i.e. March and early April and almost 12 months after blooming – the rind is sometimes affected by weather conditions, leading to slight pitting, although the tree canopy protects much of the fruit against sunburn and cold.

FORTUNE
An increasingly important mandarin in Spain

When grown in Spain and fully mature the flavour is sweet to very sweet and there is a good sugar to acid balance. However, the acid level is inherently high and harvesting must be delayed until it has dropped to acceptable levels. In some parts of the world, such as California, Israel and Swaziland, the acidity does not reach acceptably low levels before the rind has become puffy and it has been concluded there that the Fortune has few prospects. However, in Spain, where it is believed a less acidic selection of Fortune has been developed, the fruit has exceptional eating quality and achieves outstanding returns from markets in Western Europe.

Although often completely seedless, or having only one or two seeds per fruit, Fortune will set many seeds if planted near suitable pollinating varieties.

It is presently the second most popular mandarin being planted in Spain, after Nules, and accounts for over 15 per cent of all mandarins' nursery orders.

FREMONT

The original cross between Clementine and Ponkan mandarin was undertaken by P. C. Reece in Florida and further selection was done by J. R. Furr in California before it was released in 1964.

Trees of Fremont have moderate vigour and grow to a medium size, are precocious and very productive with much of the fruit borne at the ends of the branches, so that the canopy affords little protection against sunburn or frost.

Fremont is an early maturing variety and arguably the most attractive of all the mandarins, having brilliant reddish-orange colour early in the season and good shape and peel texture like the Clementine parent. It peels less readily than the Clementine, since the rind is brittle and tends to break into small pieces and is somewhat more tightly adhering. It is nevertheless a true 'easy peeler'.

Its reddish-orange internal colour matches that of the rind and it has a good Clementine-like aroma. Flavour is sweet, but even as late as one month after maturity the acid level is still fairly high. The quality never reaches that of the Clementine despite the high sugar content of the juice. The segment walls are rather tough like those of the Ponkan parent.

It has exceptional late-hanging characteristics and in Turkey by mid-March – three months or more after reaching maturity – it has not lost much quality. However, at that time its eating quality cannot compare with that of the Hernandina Clementine from Spain.

In solid-block planting Fremont will often set 15 or more seeds. Fruit size has been very small in desert areas of California, while the same shortcoming is evident in the Central Valley in years of heavy cropping. As a result, Fremont is no longer being grown in California. Only in Turkey, where it is restricted to the Adana–Mersin region, has it been planted on any scale.

IMPERIAL

(Early Imperial)

Imperial is an Australian variety, originating in 1890 at Emu Plains, west of Sydney, and believed to be a hybrid of the Mediterranean mandarin. It was introduced into Queensland by the Darrow family of Gayndah in 1956.

The trees are vigorous, of medium size and thornless and have long narrow leaves. It has a marked tendency to alternate-bearing.

The Imperial is an extremely early maturing mandarin (as early as Owari satsuma), of medium size but small when setting a large crop. It bears some resemblance to a Clementine in shape and peel texture. However, the colour is never better than yellowish-orange when grown in the Central Burnett District, Queensland. In the more southerly Murray River region with cooler autumn temperatures the colour is much improved.

The rind is thin, leathery and moderately tightly adhering but it is easily peeled and does not become puffy when left to hang on the tree for long periods. The fruit is moderately seedy, averaging about five seeds per fruit. Juice content is only moderate and the flavour is sweet but lacks richness, and at times is somewhat acidic. Production is limited to Australia where it is a popular variety because of its early maturing characteristic. Imperial is planted on almost 1,000 ha and is now the second most important mandarin after Ellendale.

FREMONT

IMPERIAL

KARA

This is an Owari satsuma–King mandarin hybrid produced in California by H. B. Frost in 1915 but not released until 1935. The tree, which is productive but tends to be alternate-bearing, has moderate vigour, an open habit like the satsuma but is larger and the branches are more drooping.

Fruit size is medium to large for a mandarin, and its shape not unlike the satsuma but sometimes very slightly necked. The rind is rough and uneven which detracts markedly from its appearance, and the areolar area is pronounced. Rind colour is deep orange and the fruit is easily peeled since the rind is often puffy before internal maturity is reached. It is susceptible to creasing, especially on rootstocks such as Troyer citrange and Swingle citrumelo. Despite very high sugar levels late in the season, the acidity often remains unacceptably high. Seed content varies from only one or two to as many as 15 per fruit.

KARA

At one time in the late 1960s Kara was planted on quite an extensive scale in Spain with the intention of taking advantage of a gap in the market for mandarins in late winter and early spring. However, consumers' reaction to its high acidity and seediness was predictably poor. The quality shortcomings mentioned above have led to many Kara trees being top-worked to other more promising varieties and it is seldom planted now.

A small quantity of Kara is produced in Australia where it is harvested in September and October. Since this is also the period in which the Murcott (Honey tangerine) matures it would be surprising if the poorer quality Kara could compete with this increasingly important variety.

An improved selection of Kara has been made by R. Bono at IVIA, Moncado, Spain, which is somewhat rounder in shape, slightly less seedy but most importantly less inclined to develop rind puffiness at maturity.

KINNOW

Kinnow has the same parentage as the Wilking, but the tree is larger, more vigorous and also shows a very strong tendency to alternate-bearing.

A late maturing variety, the fruit is of medium size for a mandarin with extremely smooth rind which, although tightly adhering, is usually easily removed but in California it is not readily peeled. The fruit hangs well on the tree without becoming puffy. Although excessively seedy, the segments are very tender and juicy, with an extremely high sugar content which to many palates is syrupy-sweet despite reasonable acidity.

Never grown commercially in the Mediterranean area, and planted on only a limited scale in Arizona and California, Kinnow found favour in Pakistan and India where its excessively sweet flavour is much appreciated.

KIYOMI

This tangor was the result of a cross between Miyagawa satsuma and Trovita orange made at Okitsu Fruit Tree Research Station, Japan. It was named and released in 1979. The trees are similar to the satsuma in their cold-hardiness.

The fruit is medium-large in size, averaging 70 to 80 mm in diameter, and is of attractive, flattish, orange-like appearance with a smooth, extremely thin but easily peeled rind. It has a particularly soft feel due to the fine satsuma-like nature of the rind and the tender texture of the segment walls and flesh. The Kiyomi has the sweet flavour of a navel orange, has deep orange coloured flesh and is both tender and juicy. However, large fruit tend to be rather insipid. An advantage is its late maturing characteristic – it is harvested from February to early April, and stores quite well.

KINNOW

KIYOMI

Kiyomi is planted on approximately 500 ha with an annual production of no more than 3,000 tons. It is not a very productive variety but the fruit quality is much appreciated in Japan. Production is mainly restricted to Saga, Kumamoto and Ehime Prefectures.

LEE

(Lee tangelo)

A sister variety to the other mandarin hybrids Osceola and Robinson, Lee resulted from a cross of Clementine with Orlando tangelo made by F. C. Gardner and J. Bellows in 1942.

The fruit resembles the Orlando tangelo in size and shape but has a deeper rind colour at maturity. However, colour development is slower than internal maturity. Harvested in Florida from late October through December, the fruit has an easily peeled, thin rind. The flesh is juicy, sweet and fairly tender although it lacks the richness of other varieties, such as Nova, which mature at the same time. Moreover, it sometimes has quite low acidity and it usually has an excessive number of seeds: often in the region of 20 per fruit. The fruit has a tendency to drop on the ground prematurely in some seasons. It has been planted to a limited extent in Florida and to a far lesser degree in Uruguay, but its long-term potential is uncertain, due to fruit quality limitations.

MALAQUINA

An Argentinian variety of unknown origin, and doubtless a hybrid, it is known by several other names, *viz*: Campeona, Bergamotto and Hybrida. The fruit is medium-large for a mandarin or a mandarin hybrid. Its appearance is similar to a very small Ugli, being flattened at the stylar-end and pointed towards the stem-end, and having a rough and very uneven rind texture. It usually attains good orange colour.

The fruit has a soft condition, and the medium thick rind is easily peeled. The flesh is orange coloured and very tender and juicy, but the flavour is rather insipid despite having fairly good sugar levels. It is invariably low in acidity and is moderately to excessively seedy. It is medium to late maturing.

MALAQUINA
An old Argentinian hybrid mandarin

In the past the Malaquina was shipped in significant quantities to Rotterdam, Netherlands, and occasionally to Canada, but its popularity has waned as improved and better quality varieties such as the Honey tangerine, Minneola and Ellendale have been developed.

MALVASIO

Believed to have originated as a chance seedling, this Argentinian variety is of unknown parentage. Tree growth is vigorous and large when fully mature.

Fruit size is medium-small, roundish in shape but slightly flattened at both apex and base, with a smooth rind which is orange in colour and thin but tightly adhering and almost as oily as Ortanique when peeled.

The fruit is juicy, and the flesh particularly tender but the flavour lacks richness despite its good sugar level. It is normally excessively seedy, regularly having 15 to 20 seeds per fruit. Its most important characteristic is its very late maturity, in August in Argentina, where it follows the Ellendale and realises premium prices. It is rarely exported and its future on the Argentine local fresh market will be challenged as more Murcott (Honey tangerine) mandarins are produced.

MICHAL

Of Israeli origin, it is believed to be a natural hybrid of Clementine and Dancy tangerine. Of good shape with smooth and deep reddish-orange rind colour, the Michal is early maturing, easily peeled like the Clementine and of good eating quality. It seems to be well adapted to the warm climatic conditions of Israel, where other Clementine selections have proved unreliable.

Sweet flavoured and with only moderate acid level, the Michal has a good aroma and is easily peeled. Unfortunately, it has three serious defects. It has proved to be strongly alternate-bearing, with wide fluctuations in yield between 'on' and 'off' years. It has a marked tendency to produce small fruit (mostly in the 50 to 65 mm range); and it has variable seed content, often averaging as many as six per fruit.

At one time it was hoped the Michal might prove successful on European markets, but against better quality Clementines from Spain and Morocco it failed to achieve a niche in the market and exports were curtailed, although it has established a position on the Israeli local market.

MALVASIO

NOVA

(Clemenvilla, Suntina)

A hybrid between the Fina Clementine and Orlando tangelo (Duncan grapefruit × Dancy tangerine) made by F. C. Gardner and J. Bellows in Florida in 1942. Although only officially released in 1964, some earlier commercial plantings were made in Florida.

The trees are vigorous, well developed and have many distinctive mandarin characteristics but are thorny. In semi-tropical areas such as Florida Nova matures early – in November – but in Spain it is later than most Clementines and is usually harvested from mid-December onwards.

The fruit is medium-large for a mandarin, comparable in size to its Orlando parent, but the rind colour is a more attractive reddish-orange. Peeling is initially less easy than the Orlando due to the firmness of the fruit, the thin rind and its tight adhesion, but once started it is easily completed with no oiling of the hands in the process. Moreover, virtually all the albedo is removed with the rind, leaving the segments as clean as the best Clementine.

Nova internal quality is extremely high. The colour is deep orange, and the segments are very juicy and tender with a fine sweet flavour, not unlike that of the Clementine. Acid levels are moderate, resulting in a high sugar to acid ratio.

Being self-incompatible the fruit is seedless when planted in isolation, and in some areas pollinating varieties such as Orlando or Temple are recommended to improve fruit set. However, like the Clementine, the presence of seeds counts heavily against the Nova with consumers who are prepared to pay premium prices only if the fruit is seedless.

While the Nova hangs well on the tree without becoming puffy, it does show a marked loss of quality with the development of severe granulation; this is especially problematic under desert conditions. For this reason harvesting should not be delayed unduly nor should this variety be propagated on a vigorous rootstock such as rough lemon. Occasionally before reaching maturity small cracks may develop in the rind at the stylar-end of the fruit. Under certain temperature and humidity conditions delayed harvesting results in an increase in 'water spot' which causes the fruit to drop to the ground. Nova is sometimes affected by sunburn.

It is currently being planted on an increasing scale in Spain and Israel where it is referred to as Clemenvilla and Suntina respectively. In Florida it is gaining in popularity with producers at the expense of the Orlando tangelo.

NOVA
Nova is seedless when grown apart from pollinating varieties

OSCEOLA

Resulting from the same cross as the Lee and Robinson (Clementine × Orlando tangelo), the Osceola is somewhat smaller in size than its sister varieties, with a slightly flatter shape.

The rind is deep orange or reddish-orange at maturity but can develop excessive puffiness soon afterwards. While peeling is easy, the rind oil sometimes has a strange and uncharacteristic aroma, not unlike that of the tomato. The flesh is well coloured and has a high sugar level, but it sometimes also has high acidity. The flavour is undistinguished and unlikely to appeal to most palates. It needs to be cross-pollinated to achieve adequate fruit set but this results in the fruit being excessively seedy: often 12 or more per fruit. The Osceola has failed in California and has been planted on only a limited scale in Florida where it matures in November.

PALAZZELLI

This variety resulted from a cross between the common Clementine and King mandarin made at the Citrus Institute, Acireale, Sicily, in 1952.

The trees are vigorous, attain medium size and are very productive but have a marked tendency to alternate-bearing. The leaves resemble those of the Clementine parent.

Palazzelli fruit is medium size and the rind has a similar colour and texture to the Clementine and is easily peeled. Its outstanding features are its good Clementine-like flavour and its exceptional late maturity which allows harvesting as late as May.

Palazzelli stores well but unfortunately it regularly sets five or more seeds per fruit even in solid-block plantings. Although a popular variety in Sicily, it is unlikely to gain much favour on overseas markets in Western Europe where the demand is for seedless mandarins.

PIXIE

The parentage of Pixie is not known with certainty other than the fact that it originated from seed open-pollinated on a King mandarin × Dancy tangerine hybrid named Kincy in California in 1927 and released in 1965.

This late maturing variety makes a vigorous, large tree with a tendency to alternate-bearing. The fruit is small and firm, more like an orange than a mandarin, and satsuma shaped with a small neck. The rind is pale orange to yellowish-orange in colour, thick and tightly adhering but easily peeled, although much albedo remains on the segments.

The segment walls are rather tough and the pulp has a distinct dry taste and lacks juice. Reports on its eating quality are conflicting: sometimes the flavour is said to be pleasantly sweet and rich, but more often it is of mediocre quality. It is almost always seedless even when grown near other varieties. It is not recommended for planting in hot desert areas of California and Arizona where it will not set fruit, nor in the cool coastal regions where quality is poor. Pixie has never reached commercial status and is regarded as being of little potential.

PIXIE

PONKAN

(Batangas, 'Chinese Honey' orange, Mohali mandarin, Nangpur suntara, Warnurco)

This is the most widely grown mandarin in the world, being common throughout south China and southern Japan, the Philippines (Batangas) and India (Nangpur suntara). It is also the most popular mandarin in Brazil, accounting for almost 50 per cent of the mandarin production in Sao Paulo State.

The Ponkan tree is vigorous and of medium size for a mandarin, having a characteristic upright growth habit. Although productive, it tends to be alternate-bearing.

The fruit is early-midseason – early November in semi-tropical areas of south China – but in central India the Nangpur suntara will flower on three occasions and set fruit. Each fruit set is quite distinct in shape and quality.

Ponkan fruit is large for a mandarin, averaging around 80 mm in diameter, and has a very oblate shape. There is usually a small neck like the Dancy tangerine, and a fairly large depressed area at the stylar-end where a small navel is often found.

The rind is moderately thick, and extremely loosely adhering, becoming puffy if left on the tree for more than a short time after reaching maturity. The rind is frequently split next to the navel, and is easily torn near the calyx when picked. It is one of the easiest mandarins to peel, and the rind oil has a distinct aroma similar to that of the Mediterranean mandarin.

Segment walls are quite tough and the juice content fairly low. In China, Ponkan is often harvested earlier than at peak maturity, when sugar content is low and full flavour has not developed. When mature it has a sweet, pleasant, low acid flavour. It lacks the richness of the Clementine and similar varieties, or the delicate, unique flavour of the satsuma, and the segment walls always remain rather tough. In India the flavour of the Nangpur suntara is reported to be good, distinctive and greatly appreciated on the local market.

A selection of Ponkan, the Oneco, was imported into Florida in 1888 and differs from it in several respects: it has rougher rind texture, is seedier and later maturing. In Brazil the Oneco is known as Cravo Tardia, and is fairly extensively planted there.

ROBINSON

From a cross between a Clementine and Orlando tangelo made by F. C. Gardner and J. Bellows in Florida in 1942, and a sister variety to Lee and Osceola, the Robinson is a very early maturing hybrid mandarin.

The virtually thornless tree has many of the Clementine characteristics, particularly leaf shape. The fruit is borne near the tips of the branches and, since the wood is brittle, there is a tendency for the limbs to break and the tree to collapse when a heavy crop is being carried. Although one of the most cold-hardy varieties, it is susceptible in Florida to twig and limb dieback.

Robinson fruit is flattish and not unlike a Murcott in shape, but there is often a small neck. Its size is similar to the Dancy tangerine. The rind is so smooth and so thin that it is slightly ridged,

PONKAN

ROBINSON

corresponding with the segments. Deep orange in colour, the rind is tightly adhering, leathery in texture and releases a lot of oil when peeled. Moreover, it tends to break into small pieces, sometimes making it messy and difficult to remove. Often much of the albedo adheres to the segments.

The pulp, although juicy, sometimes has a coarse and somewhat ricey texture. The flavour is usually sweet but without much richness, particularly when produced on a vigorous rootstock, which also makes the fruit prone to granulation. It regularly has ten or more seeds per fruit if interplanted with the recommended pollinators Temple, Lee or Orlando.

Robinson is grown on about 1,000 ha in Florida and has been evaluated elsewhere. It is unsuited to the desert areas of California and Arizona due to granulation and excessive seediness, and it seems unlikely to have much potential in other areas.

SUNBURST

Sunburst is a Florida hybrid between sister varieties, Robinson (Clementine × Orlando tangelo) and Osceola, made by P. C. Reece in 1961. It matures early (but somewhat later than the Robinson) and is harvested in Florida during a six-week period starting in early November.

One of the most impressively attractive of mandarins, the fruit has an outstanding deep reddish-orange external colour similar to that of the Minneola, and an extremely smooth and particularly thin rind – so thin that the outline of the segments can be seen as ridges on the rind surface. Although not messy to peel, the rind is so thin and brittle that it breaks into many small pieces. Segment walls are rather tough and fibrous although the pulp is not unduly lacking in juice. In solid-block plantings, the fruit sets only one to two seeds, and recommendations in Florida list Temple, Orlando, Nova or Robinson as pollinators to increase fruit set but this also results in seedy fruit. In South Africa and southern California the flavour lacks richness, and although the sugar content is quite high it is masked by noticeable acidity.

Nevertheless Sunburst realises high premiums over Robinson on the US fresh fruit market. It is unlikely to be of much potential in Mediterranean countries where far better quality Clementines are produced but it will possibly be important in areas too hot and humid for Clementines.

TANKAN

An extremely ancient variety grown extensively in south China and in the warmer regions of mainland Japan and on Okinawa, the Tankan is regarded as being a possible hybrid between Ponkan mandarin and sweet orange.

The tree is quite large and spreading with large orange-like leaves, and is very productive.

Tankan is medium-late maturing, being harvested from January to March in southern China and March and April in Japan. Its shape is almost the same as that of the Ortanique from Jamaica: flattened at the stylar-end and slightly pointed towards the stem. The rind is of a slightly pebbly texture and medium thickness, which adheres fairly well, but is easily removed without releasing an excessive amount of rind oil as is invariably the case with the Ortanique.

When mature the rind is well coloured, and the segments do not have the same toughness as the Ponkan. Pulp is deep orange in colour, very juicy and tender with a rich sweet flavour. Seed content is very variable, often having few seeds but occasionally as many as 20 per fruit.

WILKING

Like its sister variety the Kinnow, the Wilking resulted from a Willowleaf mandarin–King mandarin cross in 1915 by H. B. Frost in California and released in 1935.

The tree is moderately vigorous and of average size but has a very strong tendency to alternate-bearing. Sometimes the trees die from the very heavy load of fruit. Although the fruit is rather small it very much resembles a satsuma in appearance but has much firmer condition. It peels particularly easily and the flavour is very good with fine aroma, sweet taste and adequate acidity.

However, the larger fruit may develop granulation and all sizes are unacceptably seedy, averaging in excess of ten unusually large seeds per fruit. Medium-late maturing, Wilking initially met a good demand in the mid-winter European market and was fairly extensively planted in Spain and the Souss Valley, Morocco, but its excessive seediness was ultimately unacceptable.

Moreover, the trees provided pollen which would cross-pollinate nearby seedless Clementine trees. So important is seedlessness to European consumers that by royal decree all Wilking trees in Morocco were eradicated in the mid-1970s!

SUNBURST

WILKING

TANGORS, TANGELOS AND OTHER MANDARIN-LIKE VARIETIES

This loosely grouped collection of varieties have few features in common but are classed together since it would seem inappropriate to include them with any others.

HASSAKU

HASSAKU

Like the Natsudaidai, the Hassaku is probably a pummelo × mandarin hybrid found originally in a temple garden on a Japanese island in Hiroshima Prefecture around 1860.

Hassaku trees are vigorous, have no thorns, and the leaves are pummelo-like in shape.

The fruit is large and resembles a Cyprus or Israeli Marsh grapefruit in shape and rind texture, being slightly flattened and fairly smooth. With orange-yellow coloured rind, the fruit is firm but has a thick but very easily peeled rind. The flesh is firm, orange in colour, not very juicy, but tender. The segment walls are thick, exceedingly tough and noticeably bitter. However, when the segment walls are separated and discarded, the flesh has a good, fairly sweet, bitterness-free flavour and is only moderately acidic. Hassaku regularly has 15 to 20 large sized seeds per fruit.

Noma-Beni-Hassaku Noma-Beni-Hassaku is a reddish-orange mutant which has been extensively planted in place of the common Hassaku.

Hassaku does not hold on the tree as well as Natsudaidai and will drop readily in cold weather. It is common practice therefore to pick the fruit before the end of December. Although it stores well it cannot be kept later than the end of April without significantly affecting internal quality.

Production is on an extensive scale covering about 9,000 ha, mostly in Wakayama, Ehime and Kagoshima Prefectures and totalling approximately 200,000 tons per annum.

IYOKAN

(Iyo)

Discovered in 1883 in Yamaguchi Prefecture by M. Nakamura at Obu-gun, it was developed in Ehime Prefecture (formerly named Iyo), Japan. The Iyokan is believed to be a natural tangor.

The fruit, of which there are several selections, is medium to large compared with other tangors. It is round in shape, often slightly flattened and has a depressed area at the stem-end. It is a most handsome and attractive fruit with a smooth or very slightly pebbled rind of deep orange to reddish-orange colour, almost as intense as that of the Temple. The rind ranges from moderately thick to thick, becoming puffy with age, and is very easily removed with virtually no albedo adhering to the segments.

Segment walls are rather tough but easily separated from the firm, orange-coloured flesh which, although tender and melting in texture, is only moderately juicy.

The flavour is delicately sweet and free from bitterness and noticeably acidic – not unlike the Clementine in some respects. The common Iyokan has 15 to 20 seeds per fruit, but other recent selections have fewer:

Miyauchi Iyo Miyauchi Iyo mutant was discovered in 1952 and is now extensively grown in Ehime Prefecture.

The tree is somewhat smaller, less vigorous and rounder in shape than the standard Iyokan but it is very productive.

Although not markedly different in appearance from the regular Iyo it is much earlier maturing, has fewer seeds and the quality is regarded as being superior. About 80 per cent of Iyokan plantings are of the Miyauchi selection.

MIYAUCHI IYO (TOP)
OHTANI IYO (BOTTOM)
Both of these specimens are somewhat overmature

Ohtani Iyo Ohtani Iyo originated as a bud mutation, discovered in 1980 on a Miyauchi Iyokan tree. The fruit is much later maturing but it also differs from the standard Iyokan in appearance and internal quality.

The rind has exceptionally smooth texture and a more intense, deep reddish-orange colour than the common Iyokan, making it one of the most attractive of citrus fruits. Although the rind is also very thin it is not difficult to peel. The internal quality differs from the standard Iyokan in having higher acidity but the flavour is well balanced with good sugar content. It has only a few seeds.

All Iyokan selections lose quality, becoming dry and of rather insipid flavour, if stored too long. Most are marketed in the February to April period.

Grown mainly in Ehime Prefecture, Iyokans are produced to the same extent as Hassaku with an annual production of approximately 170,000 tons.

Iyokans are extremely popular in Japan, and are sold at premium prices; fancy grade Iyokan retail for Yen 8,500 (£38/US$68) per carton of 25 fruits!

MINNEOLA OR MINNEOLA TANGELO

(Honeybell)

Originating in 1931 from a cross of Duncan grapefruit and Dancy tangerine by W. T. Swingle, T. R. Robinson and E. M. Savage at USDA in Florida, the Minneola has now found a niche in the variety composition of many of the world's leading citrus-producing industries.

The trees are very vigorous and develop into large specimens which require sufficient room to develop fully. They are less cold-resistant than the sister variety Orlando, and are also later maturing, being regarded as a midseason variety. The tree and fruit are particularly susceptible to alternaria spot, and the tree to greening disease where this disorder is prevalent.

Large in size, the fruit has inherited some traits from both parents. The shape is round, with most fruits having a pronounced and distinctive neck which leads to its immediate recognition by consumers. The rind has not only exceptional colour, being reddish-orange, but also has a particularly fine, smooth texture.

Mature Minneolas may feel slightly soft, giving the impression the fruit is puffy, somewhat overmature and wilted; this is often not the case. The peel is thin in relation to the fruit's size, and is not difficult to remove since it is only moderately tightly adhering. However, care is needed during peeling to avoid puncturing the extremely delicate segment walls.

Minneolas have a unique, delicious and distinctive flavour, being rich (from the Dancy), tart (from the Duncan) and aromatic. It is, however, unlikely to appeal to most young children who often appreciate sweeter varieties, but many adults find the flavour much to their liking.

When grown in a solid block the fruit has few if any seeds and, like some other varieties, it is often recommended that it be interplanted with pollinating varieties such as Dancy or Clementine (Orlando being cross-incompatible), but the resulting increased seediness will only be at the expense of increased reluctance of buyers to pay premiums for seedy fruit.

It is not recommended that it be grown on vigorous rootstocks or on soils which produce inferior quality. Furthermore, during the first few years of

MINNEOLA

bearing, producers would do well to resist the temptation to market fruit which is low in sugar content or highly acidic. The same applies to the practice of some to pick and pack the Minneola when it has attained full colour but before it is fully mature; the fine eating quality can be appreciated only if consideration is given to allowing this characteristic to develop before the fruit is harvested.

The Minneola has been shown to be very adaptable and excellent quality is achieved in widely different climatic zones such as the desert areas in California and Arizona, the Mediterranean climates in Cyprus, Israel and Cape Province, South Africa, as well as the semi-tropical summer rainfall areas of Argentina, Florida and Swaziland.

MURCOTT OR HONEY TANGERINE

(Smith tangerine)

An old tangor variety most probably a hybrid of unknown parentage created by W. T. Swingle in Florida, it was first noticed as having some potential in 1913. Propagated by two unrelated growers, Charles Murcott Smith and J. Ward Smith, the same variety was known by the names Murcott and Smith. Within the USA the fruit is now officially named Honey tangerine but elsewhere, with the exception of Brazil, it is still referred to as Murcott.

Murcott trees are vigorous, bushy in shape and have willowy branches.

Fruits are mainly borne terminally where they are vulnerable to wind, sunburn and frost damage. It is very productive but with a marked tendency to alternate-bearing and, when a heavy crop is set, the branches may break and the tree collapse. Although the trees are very cold-hardy, they are susceptible to cold damage when carrying a heavy crop.

Murcott fruit is usually of medium size but small if the tree is carrying a heavy crop. It is slightly flattened at the base and apex, and the rind is smooth in texture, yellowish-orange in colour and often wind-blemished. On account of the rind's extremely thin, smooth texture which often adheres tightly to the segments, Murcott is not as easy to peel as most mandarin hybrids, but its leathery nature enables it to be peeled without much difficulty, leaving the segments free of albedo.

Segment walls are rather tough and, although the pulp is fairly tender, it does not have a rag-free texture like the Clementine and satsuma, for example. However, the juice content is very high and it has a good, sweet and distinctive, fairly rich flavour. Honey tangerines from Brazil and Florida have a very high sugar content and are so dense that they sink in water! Despite having some fine characteristics there are an excessive number of seeds, often averaging 12 per fruit and on occasions many more. Moreover, it has not been successful in the California desert due to excessive granulation. Murcott fruit is medium to late maturing and for this reason is being seriously considered in areas unsuited to Clementine production: for example in Australia where it is the third most important mandarin after Ellendale and Imperial, and in Israel where a less seedy selection is being developed. Murcott has recently been introduced into China where, like other mandarins, it is often harvested before attaining full maturity.

Most Honey tangerine production is in Florida and Brazil. Although greatly affected by recent freezes, there are around 2,000 ha grown in Florida and prices are at a premium for fresh fruit sales compared with Temples and Orlando tangelos. In Sao Paulo State, Brazil, the Honey tangerine is the third most important mandarin after Ponkan and Cravo, but nevertheless this represents an annual production of about 180,000 tons. Some are processed but the majority are sold for fresh consumption on the local market. A limited quantity is exported.

In Japan Murcott has been successfully produced as 'house mikan' in heated plastic shelters where it matures extremely late: in April and May. The appearance of the fruit is most attractive since it is blemish-free, reddish-orange in colour and hardly recognisable as the same variety produced under normal conditions in Brazil and Florida. Japanese production is estimated at 1,000 tons per annum.

MURCOTT OR HONEY TANGERINE
A tangor gaining in importance in many countries

NATSUDAIDAI OR NATSUMIKAN

(Summer orange)

The Natsudaidai was discovered about 1740 in Yamaguchi Prefecture, Japan, and the original tree is reportedly still alive! Its parentage is unknown, possibly being a hybrid of sour orange or pummelo and mandarin.

Mature trees are vigorous, with spreading branches and some thorns. The leaves are large and mandarin-like in shape and unusually dark green in colour.

The fruit is large, about the size of a grapefruit, with a flattened stylar-end. Yellowish-orange in colour, the rind is fairly rough and sometimes uneven in texture, moderately thick but easily peeled. Excessively seedy – 20 to 30 seeds are normal – and extremely bitter, it is also highly acidic for a citrus fruit which is eaten out-of-hand (2 to 3 per cent acid level).

The flesh has a somewhat coarse texture and the segment walls are very tough. A late maturing variety traditionally harvested in April and May, it is now mostly picked earlier in January and stored until early summer, during which time the acidity

KAWANO NATSUDAIDAI

falls somewhat. Nevertheless it would probably be unacceptable to the palates of those unfamiliar with its extremely bitter and highly acidic flavour. The fruit is normally peeled in the usual way but segment walls are also removed and discarded along with the seeds, leaving only the flesh to be consumed.

Several improved selections from the common Natsudaidai have been developed which have only half the acid content and are much earlier maturing. They are collectively referred to as Amanatsu.

Kawano Natsudaidai Found and named in 1950 the appearance of the Kawano Natsudaidai is more attractive than the common Natsudaidai since it is of deeper orange colour and has less coarse rind. It is regular in shape like a grapefruit but slightly more flattened. Unlike the common Natsudaidai which often develops some puffiness, Kawano Natsudaidai has firm condition, a thick very easily peeled rind, thick tough segment walls, and is somewhat earlier maturing.

It is extremely seedy with 15 to 20 seeds per fruit. If consumed whole and without removing the segment wall, the flavour is excessively bitter like the common Natsudaidai but less acidic. The flesh is light orange coloured, firm but not very juicy and when eaten in the traditional way by discarding the segment walls it has a slightly bitter but refreshing flavour with some sweetness – similar to a carbonated 'bitter orange' drink.

Other Amanatsu selections of note are: Tachibana Orange, a smooth rind mutation registered in 1974; and Beni Amanatsu, registered in 1975 and with a most attractive reddish-orange rind not unlike the Temple in colour.

Natsudaidai (particularly Amanatsu) are grown on a very wide scale in Japan, particularly in Ehime and Kumamoto Prefectures, with an annual production of around 300,000 tons from some 15,000 ha. The common Natsudaidai has lost popularity and presently accounts for only 10 per cent of total production, the remainder coming from the Kawano Natsudaidai selection which commands significant premiums on the fresh fruit market.

ORLANDO OR ORLANDO TANGELO

(Lake tangelo)

Like its sister varieties Minneola and Seminole, the Orlando was developed from a Duncan grapefruit–Dancy tangerine cross in 1931.

Tree growth is similar to the Minneola although not so vigorous or large. The leaves are characteristically cup-shaped.

The fruit is early maturing, medium to large in size although smaller than the Minneola, almost round in shape and only very slightly oblate. The rind is slightly pebbly towards the stem-end, light orange in colour, very thin and moderately tightly adhering. The Orlando is not an easy fruit to peel; in fact it is almost as difficult and oily to peel as the Ortanique. Tender and very juicy, the pale orange flesh has a sweet to very sweet but rather insipid flavour with low acidity. It is often interplanted with other varieties such as Dancy and Temple (Minneola is cross-incompatible) which improves fruit set but also increases seediness. Orlando tangelos are fairly seedy and regularly have ten or more seeds per fruit.

Initially more popular than the Minneola but largely restricted to its native Florida, Orlando was

ORLANDO

also produced in Jamaica under its original name Lake tangelo and in California and Israel, but performed disappointingly in these latter areas. Its cold-hardiness during the 1962–3 Florida freeze encouraged a planting boom and by 1972–3 the Orlando was overproduced with a crop of more than 500,000 tons, much of which had to be processed for juice, and prices became depressed. Successive recent freezes have seen a further dramatic decline in the importance of the Orlando tangelo and it is being replaced with varieties of better quality such as Nova.

ORTANIQUE

(Topaz, Tambor, Mandora)

Discovered in Jamaica (as were the other natural hybrids Temple and Ugli) the Ortanique is a tangor first propagated by C. P. Jackson of Mandeville in 1920; its name being a synthesis of *OR* (-ange), *TAN* (-gerine) and (un-) *IQUE*.

The tree is very vigorous, spreading and reaches a large size. It is a most reliable and productive bearer, the fruit maturing late in the season at about the same time as the Valencia. It can be left

ORTANIQUE
This shows the effect of climatic conditions on the appearance and quality of the Ortanique grown in Jamaica (left), Israel (Topaz, centre) and Cyprus (Mandora, right)

on the tree for a reasonable length of time without becoming puffy or losing quality.

The fruit is of medium size, and slightly flattened at the stylar-end where a small navel is often formed. The shape, peel texture and thickness, as well as external colour and internal quality, are affected greatly by the area in which the fruit is produced.

Until the early 1970s production was very limited and restricted to Jamaica, but the Ortanique has shown good adaptability to less tropical climates. It is now grown on a modest scale in neighbouring Honduras (known there as Ormanda), Australia (Australique), Cyprus (Mandora), Israel (Topaz and Tangor), Swaziland and South Africa (Tambor), the plethora of names brought about by a trademark on the original 'Ortanique'.

Under Jamaican tropical conditions the rind is smooth, pale orange in colour and extremely thin. As the climate in which it is produced becomes more semi-tropical and Mediterranean, the rind becomes progressively coarser, deeper orange and somewhat thicker although never more than of medium thickness.

The rind is leathery and initially most difficult to remove and this is never accomplished without a great deal of rind oil being released. However, despite having somewhat tough segment walls, the pulp is tender and extremely juicy, often exceeding 60 per cent by weight of the whole fruit.

The juice is of outstanding colour, especially from origins other than those with tropical climate, while the flavour is extremely sweet but well balanced with acidity and has a strong, rich aroma. It does not develop delayed bitterness.

Seediness is variable: from Jamaica there are few if any seeds, while from Cyprus as many as ten seeds per fruit are sometimes present. There is a tendency for the rind to split near the navel before the fruit is fully mature. Oleocellosis is sometimes problematic if due care is not exercised during picking and packing. In addition to hanging well on the tree, the Ortanique may be successfully stored for a considerable period without the juice developing off-flavours.

PAGE

This was created in Florida by P. C. Reece and F. C. Gardner in 1942 from a Minneola (Duncan × Dancy) and Clementine cross, and released in 1963.

The tree is moderately vigorous for a mandarin hybrid, is thornless and extremely productive: so much so that the branches often need support to prevent breaking under the weight of fruit. There is often a noticeable leaf drop in the winter period. Page is reported in Australia as being very sensitive to the herbicide Bromacil (Hyvar).

Usually referred to as an orange, Page has no particular external characteristics to suggest it is anything other than a well-coloured, round, smooth, small orange variety. Although the rind is fairly thin and easily peeled, it is excessively oily – every bit as oily as an Ortanique – and often the flavour is not in any way more than simply sweet and pleasant. However, in Florida, it is reported to have a distinctively rich flavour – almost the best of the orange varieties.

Without a pollinator such as the Dancy, Orlando or Valencia recommended in Florida, the Page bears a light crop and when pollinated the fruit is small and excessively seedy, averaging about 12 seeds per fruit.

PAGE

Page reaches maturity early in the season and holds well on the tree. At one time it was regarded as having some commercial potential, but there are insufficient good characteristics to justify such optimism.

SEMINOLE

The third sister variety with Minneola and Orlando, the Seminole has never gained anything like the same popularity except in Japan and to a small extent in New Zealand.

The tree attains the same size and vigour as the Minneola, the leaves are cupped like the Orlando and the fruit is borne inside the canopy.

Deep reddish-orange like a Minneola in colour, the fruit is medium sized, round in shape and with a thin but rather pebbly rind (due to creasing) which is moderately tightly adhering like the Orlando tangelo but easily peeled. Tender and extremely juicy, the Seminole is late maturing, almost invariably having a high acid level which many find unacceptable. However, the average Japanese palate is accustomed to the highly acidic flavour of some of its own citrus varieties, and it is not surprising the Seminole is cultivated there on a modest scale of about 500 ha, producing an estimated 6,000 tons. It is grown principally in the warmer areas of Honshu, Kyushu and Shikoku, but production is on the wane in Japan. It is reported as having good storage characteristics.

In the marginal citrus climate of New Zealand's North Island, where it is referred to simply as Tangelo, the flavour of the fully mature fruit, even when grown on trifoliate orange rootstock, is not excessively acid. It closely resembles the Minneola in flavour and flesh texture, but is somewhat more seedy. An estimated 4,000 tons are produced on 500 ha. There is a small export trade in Seminoles to Japan. In New Zealand Seminole is a popular 'backyard' variety.

TEMPLE OR TEMPLE ORANGE

(Royal mandarin)

Of unknown origin but almost certainly a mandarin–orange hybrid found growing wild in Jamaica in 1896, it was propagated on a limited scale until released by Buckeye Nurseries, Winter Park, Florida, in 1919, named after a former manager of the Florida Citrus Exchange, W. C. Temple.

SEMINOLE

TEMPLE

The trees are of medium vigour, bushy, somewhat thorny, and have mandarin-like leaves. They are productive but are sensitive to cold damage, with the fruit deteriorating on the tree or dropping from it following freezing weather. It is also most susceptible to citrus scab.

It is a midseason variety maturing shortly after the Minneola and Ellendale.

Temple fruit is of medium-large size and its shape is slightly flattened at the stylar-end, with the rind evenly pebbled and slightly furrowed at the stem-end. It is a handsome, highly coloured fruit, having the deepest reddish-orange colour of any citrus variety. The rind is fairly thin and easy to peel without being oily. The rind oil has a distinctive aroma. Pulp colour does not match that of the rind, being light orange rather than deep orange. Flavour is good, with high sugars, moderate acidity and a unique and subtle flavour described as 'spicy'. The segment walls are slightly tough but the flesh is fairly tender. Temple is self-fertile and typically has 15 to 20 seeds per fruit. However, two selections with much lower seed counts have been developed: Sue Linda, with about eight seeds per fruit, and Thoro which is often seedless and averages only one or two seeds per fruit. Sue Linda also has smoother rind texture, a slightly higher sugar to acid ratio and better juice colour.

Temple fruit quality is influenced markedly by tree nutrition; it must be grown on a good rootstock and adequately fertilised if good quality is to be obtained.

At one time Temple was produced on 10,000 ha in Florida and the crop totalled 170,000 tons, but following recent freezes it now stands at only half that quantity. Approximately 1.7 million cartons are sold on the US fresh fruit market.

Outside Florida the Temple has never achieved any great significance. It is grown on a very limited scale in California, being restricted there to the hottest desert areas; elsewhere the fruit is too acidic. Californian Temples are marketed as 'Royal' mandarin.

In Israel, where a small area was planted, it has been found to be unprofitable when competing against better quality mandarin varieties on the European market.

THE GRAPEFRUIT

Citrus paradisi

The origin of the grapefruit is uncertain but recent studies suggest it is a natural cross between sweet orange and pummelo (or Shaddock) which occurred in the 1700s on Barbados in the West Indies. It is believed pummelo seeds were taken to Barbados by an Englishman, Captain Philip Shaddock, around 1649, but it was not until 1823 that mention of the grapefruit was first recorded. Known locally as 'the forbidden fruit', it was given the species name *Citrus paradisi* in 1830.

Of equal uncertainty is the origin of the name Grapefruit: some suggest the flavour resembles that of the grape (some grape!), while others believe it is the way the fruits are borne in small clusters on the tree, rather than individually as with pummelos.

Grapefruit was introduced to Florida from the Bahama Islands by a Frenchman, Count Odette Phillippe, who settled in Safety Harbor, Tampa Bay, in 1823. However, it was not until more than 60 years later in 1885 that commercial consignments were first shipped to the northern cities of New York and Philadelphia.

Since then production has steadily increased until today around 3.8 million tons are produced worldwide. The United States is the world's leading producer with 2.3 million tons, about 45 per cent of which is shipped fresh. Other important producers are Argentina, Cuba, Cyprus, Israel, Mexico and southern Africa (Swaziland, Mozambique and South Africa).

MARSH GRAPEFRUIT SHOWING TYPICAL BEARING HABIT WHICH SOME BELIEVE RESEMBLES THAT OF A BUNCH OF GRAPES

Grapefruit is known by different names:

Pamplemousse	French
Pompelmo	Italian
Pomelo	Spanish and others

The quality of grapefruit, like that of most citrus varieties, is influenced by several factors, particularly climate and rootstock. The sweetest, juiciest and most bitter-free fruit is grown in semi-tropical summer rainfall regions such as Texas, Florida and the lowveld areas in southern Africa. In the cooler, drier Mediterranean climates there is a tendency for the fruit to have a thicker rind, lower sugar and high acid levels in the juice and to have some bitterness. The best quality grapefruit is grown on sour orange, citrange and Swingle citrumelo rootstocks; vigorous rootstocks such as rough lemon invariably produce fruit of inferior quality.

Grapefruit varieties may be conveniently divided into two groups. The first comprises white or common varieties of which only Marsh is of any significance. The second are the pigmented varieties which are increasing in number and in popularity with consumers.

WHITE OR COMMON VARIETIES

While there have been many white grapefruit varieties identified, only two are grown on any scale, of which Marsh predominates.

DUNCAN

The origin of the Duncan can be traced back directly to the first grapefruit introduced to Florida. The original Duncan tree was planted around 1830 near Safety Harbor, Florida, but it was not until 1892 that it was named and propagated by A. L. Duncan of Dunedin. All other grapefruit varieties have arisen from the Duncan variety.

Duncan trees are vigorous, large and very productive, while the fruit is larger than Marsh and the tree more cold-hardy.

Unsurpassed for flavour, the flesh of Duncan is firm but very juicy, has good acidity but is also high in sugars giving a well-balanced, rich, sweet

DUNCAN
The finest flavoured grapefruit variety

MARSH
The predominant white grapefruit variety

and pronounced flavour. It reaches maturity in Florida in November but is at its best in February and March when acidity is lower.

Unfortunately, it is excessively seedy, having typically 30 to 50 seeds per fruit. However, it is still regarded as the standard in quality of all grapefruit varieties. Seediness is of lesser importance to processors who still regard Duncan as being the most suitable variety on account of its segment stability and flavour. It is also the preferred variety for juice production.

Planted almost exclusively in Florida and on a greatly reduced scale of late, Duncan is virtually all processed and accounts for only 7 per cent of Florida's grapefruit production.

MARSH

(Marsh Seedless)

Originating as a seedling from a Duncan tree in Lakeland, Florida, around 1860, it was eventually propagated and named in 1892 by C. M. Marsh: Marsh Seedless or more commonly simply Marsh.

Like the Duncan the tree is vigorous and very productive, attaining large size although regarded as being more cold-sensitive. Marsh fruit is slightly smaller than Duncan but far less seedy, having typically just two or three per fruit, but is rarely ever truly seedless.

The juice sacs are smaller and the pulp has a somewhat smoother and lighter appearance than Duncan. It has very slightly thicker rind but a high juice content of sweet flavour that is of rather high acidity early in its season. While both varieties may reach minimum maturity standards at about the same time in November, Marsh may hang later than Duncan, although towards the end of the season the acidity is low and the flavour rather insipid.

Marsh is far and away the most predominant variety worldwide, although within Florida it is rivalled in popularity by the Pink seedless varieties, which collectively account now for about 38 per cent of production, while in Texas production is given over almost entirely to pigmented varieties. Marsh processes well for both the juice and segment markets.

PIGMENTED VARIETIES

Pigmented grapefruit varieties derive their colour from the pigment lycopene, unlike pigmented oranges in which anthocyanins are responsible for the distinctive colour. Moreover, the climatic conditions which bring about the expression of pigmentation in the grapefruit are different from those with blood oranges. Whereas anthocyanins develop during the autumn and winter months when temperatures fall to quite low figures, lycopene production is only well achieved with prolonged high temperatures. The result is invariably that pigmented grapefruit show best colour development under growing conditions similar to those in Florida and Texas.

During the last two decades the popularity of pigmented grapefruit has increased markedly in many countries but somewhat surprisingly not so in Japan. Today almost 40 per cent of Florida's production is given over to pigmented varieties and the trend is set to continue. In Texas, where pink and red varieties have long predominated, the less well coloured varieties (Ruby Red, Henderson and Ray Ruby) are marketed as 'Ruby-Sweet' and the better pigmented Star Ruby and Rio Red as 'Rio Star'. The Israeli crop of Star Ruby is packed under the name 'Sunrise', and the Redblush variety as 'Yarden Red'. Redblush from southern Africa is marketed as 'Rosé'.

BURGUNDY

The parent tree of this variety was discovered on H. J. McReynolds' property, Orlando, Florida, in 1948 and named in 1954. It probably originated from Pink Marsh (Thompson).

Planted on a limited scale, mostly in the Indian River area of Florida, it is a late maturing variety which hangs well into early summer without loss of quality.

The rind is smooth but without a trace of red pigmentation, although internally it has flesh colour almost as intense as that of Star Ruby but often

BURGUNDY

with a brown tinge. Seeds are few – just one or two – while the flesh is firm, very juicy, with a sweet flavour free from bitterness. However, its overall internal quality is regarded as being somewhat inferior to most other pigmented varieties and it is rarely planted today.

FLAME

This very recent variety originated from seed collected by W. K. Wutscher in Texas in 1973 from the Henderson variety. From the resulting seedlings, C. J. Hearn at USDA, Orlando, Florida, selected Flame in the early 1980s and budwood was released in 1987.

At present, nothing is known about the tree's characteristics with regard to ultimate size and productivity but early indications suggest the fruit has external blush similar to that of Ray Ruby but less than that of Star Ruby, while internal pigmentation is better than Ray Ruby and nearly as good as that of Star Ruby.

Although never grown in the same location as Rio Red, if Ray Ruby is used as a standard Flame has as good or perhaps better internal colour than Rio Red. Flame is seedless and is of similar quality to Ray Ruby.

Like Rio Red, Flame merits evaluation wherever deeply pigmented grapefruit is being considered as a replacement for Star Ruby.

HENDERSON

Henderson was discovered as a bud sport in 1973 on S. Henderson's property near Edinburgh, Texas, on a Fawcett Red (Everhard Red) tree which was a little-known variety itself originating as a bud mutation on a Pink Marsh tree near Alamo in 1940.

There are no differences between Henderson and Ruby in time of maturity or any of the juice quality characteristics. However, Henderson has better pigmentation both externally and internally, and importantly, flesh colour is retained much better with age than Ruby.

RAY RUBY

The two original trees are of uncertain origin and were discovered in 1970 on the property of R. Ray near Mission, Texas. Although the trees are indistinguishable from those of Ruby, the fruit has a redder blush and overall redder colour internally, but the differences are not apparent until mid-December. Thereafter rind colour approaches but never quite equals that of Star Ruby.

Slightly sweeter than Ruby and with better internal colour, it is regarded as a superior variety and the juice makes a more attractive product when processed.

RIO RED

This is the most recently introduced pigmented variety which, like Star Ruby, was developed by R. A. Hensz in Texas from budwood of a Ruby Red seedling. Following irradiation in 1963, a tree with fruit of a more intense red flesh colour, designated A & I 1–48, was propagated. In 1976 it produced a bud mutation subsequently named Rio Red.

Flesh colour of Rio Red is five times that of Ruby and twice that of Ray, and almost as intense

RUBY

as Star Ruby, while the rind colour is similar to Ray Ruby, both being far better than Ruby Red. In all other respects the tree and the fruit quality characteristics of the Ruby, Ray and Rio varieties are similar.

Rio Red was released in 1988 and, like other promising varieties such as Henderson and Ray, is undergoing long-term evaluation in Texas and elsewhere. It seems destined to replace Star Ruby as the principal deeply pigmented variety.

RUBY

(Ruby Red, Redblush, Henninger)

These varieties originated in the same area of Texas at about the same time and are thought by most authorities to be one and the same variety.

Ruby was discovered as a bud mutation of Thompson (Pink Marsh) in 1926 by A. E. Henninger at McAllen, Texas, and was patented and named in 1934.

Ruby matures at about the same time as Thompson but has better internal quality as well as better internal and rind pigmentation. This was the first pink grapefruit that could be immediately recognised externally without having to be cut open, and one which also had better colour than Thompson. Apart from the colour, Ruby is virtually identical to Marsh in most other fruit characteristics and typically has few seeds.

Ruby is presently the most widely grown pigmented variety worldwide, but no doubt will one day be superseded by more recent selections such as Rio Red and others.

STAR RUBY

This variety was produced by irradiating seed from the Hudson variety by R. A. Hensz, Texas A & I University, Weslaco, Texas, in 1959. Flesh of the Star Ruby is slightly redder than Hudson and it remains the most heavily pigmented grapefruit yet developed. Furthermore, external pigmentation was also enhanced and is unsurpassed by more recent selections. Although the internal colour fades during the season, it remains outstandingly strongly pigmented until the end of the season.

In addition to these two good and important characteristics, Star Ruby is almost invariably seedless, rarely having more than one or two seeds in a minority of fruits. It has an extremely thin rind, a very high juice content and a sweeter, less bitter

STAR RUBY

flavour than Marsh and other pigmented varieties. It is the standard pigmented grapefruit against which all others are measured.

However, it is evident that the irradiation had deleterious as well as beneficial effects on the genetic make-up of the variety, since it has been found under many conditions worldwide to be the most problematic of all grapefruit trees to grow. It is slow growing and develops a rather compact, stunted, bushy growth habit. In addition, it is extremely susceptible to foot rot and is herbicide-sensitive, as well as developing stem-pitting disease (CTV) at a far earlier age than Marsh and Ruby. It exhibits excessive sunburn-sensitivity in hot desert areas. Whereas most grapefruit varieties may successfully be stored for several months, Star Ruby is particularly prone to *Diplodia* stem-end rot after no more than a few weeks. Because Star Ruby trees are often lacking in vigour, fruit size is affected, with the result that small fruit predominate. This is a serious disadvantage since the significant premiums paid for this and other well-pigmented grapefruit varieties are restricted almost entirely to large size fruit.

THOMPSON OR PINK MARSH

This variety was the first seedless pigmented selection, originating as a bud mutation on a Marsh Seedless tree on W. B. Thompson's property, Oneco, Florida, in 1913. It was not named and propagated until 1924.

It has the same tree and fruit characteristics as Marsh, differing in only two respects: it matures a little earlier, and the flesh is slightly pink – and limited to the areas near the segment walls, tending to fade out with age.

Comparative Pigmentation Rating of Current Pigmented Varieties

Some varieties have been released so recently that a side-by-side comparison of current pigmented varieties has not yet been undertaken. However, a rating of each has been attempted below with regard to external and internal pigmentation in an endeavour to compare one against another.

	Rind rating	*Flesh rating*
Burgundy	Nil	****
Thompson (Pink Marsh)	Nil	*
Ruby (Ruby Red, Redblush)	**	**
Henderson	****	***
Ray Ruby	****	***
Rio Red	****	****
Flame	****	****
Star Ruby	*****	*****

Pigmentation * = poor ***** = intense

Tracing the Development of Present-day Grapefruit Varieties

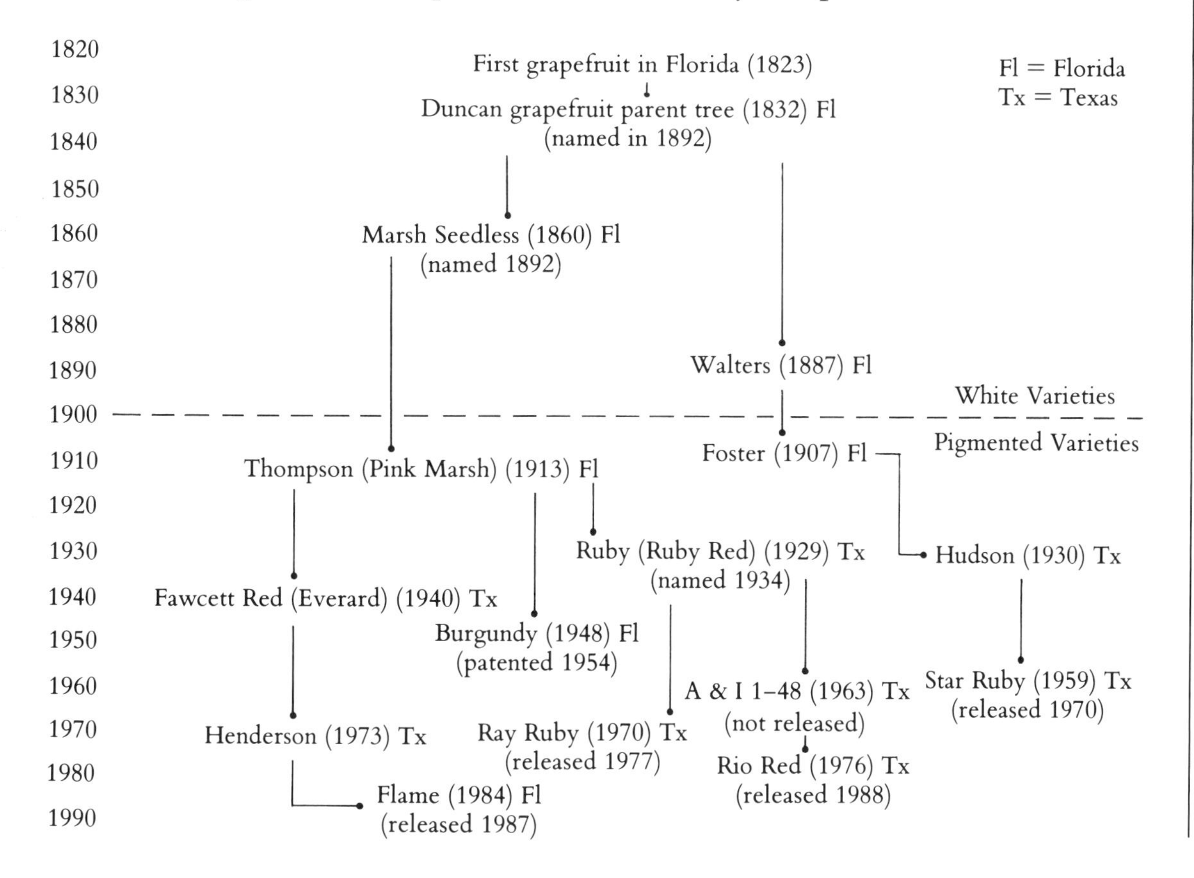

THE PUMMELO

Citrus grandis

The pummelo (sometimes referred to as the Shaddock) is the most tropical of citrus fruits and until recently was almost unknown in North America and many European countries.
The pummelo is called by many names:

Pummelo, Shaddock	English
Pamplemousse	French
Adamsapfel	German
Pompelmo	Italian
Toronja	Spanish
Limau	Indonesian
Somo, Mao	Thai
Buntan	Japanese
Yu	Chinese

Originating most probably in southern China where it is widely grown, it has been distributed throughout south-east Asia where many varieties have been developed. It is convenient to group pummelos into three types according to where they were developed: Thai, Chinese and Indonesian.

Pummelo trees vary in size from comparatively small to some which are among the largest of citrus trees; most are often drooping in habit. The leaves are larger than those of other citrus varieties and the twigs are thorny on young trees but less so when older. Pummelo flowers are large, 3 to 7 cm in diameter, and they occur singly or in small clusters.

When first encountered it is easy to see that pummelos and grapefruit are closely related but

PUMMELO TREES PLANTED ON RAISED BEDS IN THE BANGKOK DELTA PLAIN, CENTRAL LOWLANDS, THAILAND their eating quality as well as other characteristics are quite different and distinct.

There are enormous variations in fruit size, as well as shape, rind thickness, external and internal colour, but pummelos tend to be large or very large and may measure from 10 to as much as 25 cm in diameter and weigh from 500 to 1,500 g, but fruits 60 cm in diameter and weighing as much as 10 kg have been reported.

Fruit varies from round to pear shape, peel texture from extremely smooth to pebbly, but all have a surface characterised by greenish dots which are the rind oil glands. Sometimes the oil cells protrude from the peel. Rind colour is still often green when the fruit reaches maturity but, depending on the climate in which they are grown, this changes with age to light green, yellow or bronzy-yellow.

Most pummelos have a thick rind – often from 10 to 30 mm depending upon fruit size – which is soft and easily peeled, although invariably a considerable amount of albedo adheres to the segments; the albedo may be white or slightly pink in colour. Unlike a grapefruit except when it is very overmature, the pummelo's central core is usually open but in the better varieties it is closed to some extent.

The flesh colour of pummelos varies like that of grapefruit from white through pink to deep red. While most grapefruit have about 12 regular sized segments, pummelos commonly have from 16 to 18, some of which are large and extend close to the edge of the rind, but a minority are much smaller and are 'blind', being surrounded by the larger regular ones.

Whereas the vesicles or juice sacs of most other citrus varieties are loosely fused together, those in the pummelo are separate although tightly arranged and the segment walls often burst near the core before the segments are separated.

Pummelo segments are not normally eaten whole like those of oranges or mandarins, nor is it common to remove the flesh with a spoon as with grapefruit. Instead, the usual practice is to peel away the segment walls on two sides, which is easily achieved without releasing the juice or soiling the hands. An alternative way of preparing the fruit for table use is to shell out the vesicles, or 'fillet' them from the segment walls, into a dish.

Like the physical characteristics, the flavour of pummelo is far more variable than that of the grapefruit and ranges from insipid to highly acidic. Most have a pleasant, distinctly individual flavour very much appreciated by people of all ages throughout the Far East. Pummelos store better than any other citrus variety and can be held in unsophisticated facilities for many months without the need for refrigeration. Throughout China, for example, pummelos are commonly marketed year round.

The pummelo is self-incompatible but when grown in close proximity to other pummelo varieties or other citrus fruits with viable pollen, the fruit can be excessively seedy with as many as 150 large seeds per fruit. However, when planted in isolation or in a solid block most varieties produce few if any seeds.

Because pummelos hybridise very readily, the number of separate and distinctly different varieties is enormous, but the principal ones are as detailed in the following pages.

Thai Group

In Thailand pummelo trees will set several crops each year where flowering flushes follow the onset of periodic rainy seasons.

The fruit is variable in shape but usually smaller in size than those in the other groups and they are generally regarded as being of better quality.

CHANDLER

This is a hybrid pummelo made by controlled pollination of two parents, namely Siamese Sweet by Siamese Pink, undertaken by the University of California, Riverside, and released in 1961.

The tree is vigorous with an open drooping habit. Chandler fruit is early maturing, medium in size ranging from 400 to 1,000 g, and round in shape. The rind is smooth and glossy in texture, moderately thick, tightly adhering, and yellow in colour, usually with a slight blush.

The flesh is pink to sometimes red and the texture is often somewhat coarse. It is only moderately juicy but very sweet and has adequate acidity to give an overall pleasant flavour. It is, however, more acidic than the pummelos grown in the Far East or the Goliath variety in Israel.

CHANDLER

When grown in isolation Chandler has been found to be virtually seedless. It may be planted in proximity to navels, satsuma mandarins or Oroblanco without the risk of increasing seediness, since these varieties are essentially pollen-sterile. Chandler is not recommended for planting in desert areas or the cooler coastal regions of California, the former location producing coarse textured rind and the latter being insufficiently warm to give good internal colour and quality.

KAO PANNE

(Kao Pan)

This variety is of comparatively recent origin – less than 100 years! – and is grown primarily in the Bangkok area. Trees are of medium size, round-topped with a spreading habit and moderately productive.

The fruit is slightly flattened in shape and averages a small 10 to 12 cm diameter, while the rind is smooth, yellow in colour and moderately thick. The segment walls are thick, the central core small and usually solid and the vesicles are large, white and moderately juicy. Kao Panne is early maturing and the flavour is very sweet, but with a reasonable amount of acidity – although lower than that of most grapefruit.

KAO PHUANG

(Siam, Siamese, White Tassel)

This is also widely grown in the Bangkok area and exported to neighbouring countries where it is known as the 'Siam', 'Siamese' or 'White Tassel' pummelo. Kao Phuang trees are more vigorous, larger and more productive than the Kao Panne.

Fruit is pear shaped, with a long distinct neck. It is medium sized for a pummelo, with rind of greenish-yellow colour, smooth and shiny, less thick and less tightly adhering than Kao Panne, but with similar flesh characteristics. The flavour is somewhat more acidic. It is a midseason variety and holds well on the tree without loss of quality.

Some experts believe Kao Phuang has better potential than Kao Panne in other citrus regions outside the Far East since it has better vigour, is more productive and has superior flavour. However, because Kao Panne has a smaller and therefore possibly more acceptable fruit size, both varieties are well worthy of evaluation elsewhere.

The Nakon (Nakorn) has sometimes been considered identical to Kao Panne and Kao Phuang but it is now regarded as a separate and new variety, about which little has been reported.

Chinese Group

There are countless pummelo varieties produced in China, of which Mato and Shatinyu are known to me. Also included in this group is Goliath, a variety produced in Israel.

GOLIATH
A pummelo variety being produced on an increasing scale in Israel

GOLIATH

Until very recently pummelo production was restricted to tropical or semi-tropical regions, and it was generally believed that even in the warmer regions of the Mediterranean basin there were insufficient heat units for the development of good quality fruit. However, an old-established variety of unknown origin, given the name of Goliath, has been propagated in Israel from trees growing in the coastal area and planted in the Jordan Valley during the past few years.

The trees have good vigour but only moderate size and a somewhat spreading and drooping habit. They are fairly productive, bearing much of the crop inside the canopy and near to the ground. Early maturing, the fruit is medium in size, often pear shaped and sometimes has a pronounced neck. The rind is slightly pebbled in texture with protruding and conspicuous oil glands, and greenish-yellow in colour, turning to yellow later in the season. The fruit stores exceptionally well.

The peel is moderately thick but soft and easily peeled. The central core is usually open but the fruit is seedless with slightly coarse, juicy flesh which becomes more tender with storage. White in colour, this variety has a good flavour that is pleasantly sweet and lacking any noticeable acidity or bitterness.

Production in Israel has increased significantly in the recent past, with exports presently standing at around 400,000 10 kg trays. Unfortunately it is being marketed under the misnomer, 'Pomelo', this being the Spanish word for grapefruit.

MATO

(Madowendan, Mato Buntan)

Grown principally in Fujian Province, China, and nearby Taiwan and on a limited scale in south-western Japan, the tree is comparatively small in size, round in shape and drooping in habit.

Early maturing, the fruit size is medium to large in China but smaller in Japan, averaging around 750 g, and has a distinctive shape with a broad, flat stylar-end and a pronounced and tapering 'sheepnose' stem-end.

The rind has an evenly pebbly texture with large protruding oil cells. Light greenish-yellow to yellow in colour, the rind is of medium thickness while the flesh is greenish-white in colour, rather coarse in texture and only moderately juicy. The flavour is sweet but with some acidity and slightly bitter.

Shatinyu

Grown extensively in the coastal provinces of south-east China, and more recently in the Red Basin region of Sichuan Province, the tree is vigorous and attains large size with a marked drooping habit. The fruit is of medium to large size, elongated with a very pronounced neck.

Shatinyu has a moderately thick rind of pale greenish-orange turning to yellow or orange-yellow when stored. While the flesh is white, fairly smooth in texture, but juicy and tender, it has very low acidity. It has a sweet to very sweet and most pleasant flavour.

Indonesian Group

Widely distributed throughout Indonesia, Malaysia and India, as well as in Taiwan and Japan, the Indonesian group pummelos are extremely variable in quality and other characteristics but are always larger than varieties in the Thai and Chinese groups. They are usually round in shape and include:

Banpeiyu

(Pai yau)

Of Malaysian origin, Banpeiyu is the most commonly grown pummelo in Japan, particularly in the Yatsushiro area of Kumamoto Prefecture, and second only to Mato in Taiwan. The tree is very vigorous, spreading and attains large size.

The fruit is large and almost perfectly round in shape, while the rind is thick, tightly adhering and yellowish-orange when grown in Japan. It is medium-late maturing, has white flesh not unlike a Marsh grapefruit, a smooth tender texture, and is moderately juicy with a sweet flavour but sufficient acidity to give a good balance.

Banpeiyu

Djeroek Deleema Kopjor

(Hybrid 202)

This new variety – possibly of hybrid origin from the Indonesian Djeroek Deleema Kopjor – was selected in the late 1970s by H. de Lange, CSFRI, and has been developed by S. Burdette, Outspan Citrus Centre, Nelspruit, South Africa, in the past decade.

Little is known yet about the tree's vigour and ultimate size and productivity, but early indications are encouraging.

Smaller in size than other pummelo selections, the fruit is round to slightly oval in shape, with extremely smooth rind texture, like the best Florida and Texas grapefruit. The rind is slightly yellowish-green to yellow in colour, without any blush, and extremely thin. The flesh is pink – sometimes more intense in colour adjacent to the segment walls – extremely tender, very juicy and far less coarse and ricey in texture than Chandler or Goliath. Significantly, there are fewer segments than these two varieties – usually only 12 or 13 – most of which are regular in shape. It is seedless when grown in isolation.

DJEROEK DELEEMA KOPJOR
(Hybrid 202)

The flavour early in the season is mildly acidic combined with very high sugar levels, giving a well-balanced, delicious taste. Later the acidity falls and the sweetness increases to a sugar to acid ratio of around 16:1 to make this an outstanding variety in terms of flavour as well as having the most attractive appearance and acceptable size.

Preliminary indications suggest Hybrid 202 has good climatic adaptability with regard to fruit quality.

The variety name Cameron has been suggested for the Djeroek Deleema Kopjor (Hybrid 202).

JAVA SHADDOCK

Of unknown origin, but possibly from seed introduced before 1920 to South Africa, the Java Shaddock was selected by L. von Broembsen in 1970 and propagated by F. Esselen at Malelane, Eastern Transvaal.

Java trees are vigorous and attain large size with an open habit.

The fruit is large to very large with most in the range 135 to 150 mm in diameter and varies from round to slightly oval in shape. The rind is very smooth in texture and glossy in appearance, yellow in colour without any blush and of only medium thickness compared with other varieties such as Goliath, Chandler and Reinking.

The flesh is pink, almost as deep in colour as Chandler, as tender and more juicy with a very good sweet, moderately acid (for pummelo) flavour. Normally seedless, the segments are easily peeled to enable it to be eaten in the traditional way.

Unfortunately, its size is too large for the export trade since the cost of freight to Europe makes the price per piece of fruit very expensive.

PANDAN BENER

This variety, from the Batavia district of Java, has large fruit size, red flesh and good flavour but the tree is not very productive.

PANDAN WANGI

Also from Java and having similar fruit characteristics to Pandan Bener but better productivity.

Other Pummelo and Grapefruit-like Varieties

This is a very loose grouping of different varieties that are of unknown origin or have been created by citrus breeders.

MELOGOLD

Melogold (and Oroblanco) originated from a cross between an acidless pummelo and a white seedy grapefruit made by R. K. Soost and J. W. Cameron at the University of California, Riverside.

Although fruit shape is not dissimilar to Marsh and Oroblanco except for a slightly tapering neck (sheepnose) – a feature often taken as an indication of poorer internal quality – Melogold is very much larger in size, often averaging 700 g compared with 520 g and 450 g for Oroblanco and Marsh respectively. There may therefore be the serious problem with utilisation of a variety which cannot fit into the normal pattern as a breakfast fruit. Unlike the pummelo, the segments of Melogold cannot be peeled and separated from their walls without the process becoming very messy and impractical. However, in California as Melogold trees become older, fruit size is reduced and is then not regarded as being excessively large. On the other hand, Oroblanco fruit tends to become too small. As a result of this, and the more attractive darker yellow rind colour, Melogold is proving more popular with producers than Oroblanco, despite the fact that Oroblanco quality is often regarded as the better of the two varieties.

Melogold rind is smooth but develops colour slowly like Oroblanco, only achieving full yellow colour several weeks after being internally mature. Juice percentage is better than Oroblanco and similar to Marsh but, like Marsh, it is often slightly bitter very early and very late in the season. Since it is a triploid, it is usually seedless whether or not it is pollinated. The flavour of Melogold is sweeter than Marsh and different from Oroblanco, being more pummelo-like than either.

The Plant Patent rights to Melogold and Oroblanco are held by the Regents of the University of California.

NEW ZEALAND GRAPEFRUIT OR POORMAN ORANGE
A slightly overmature specimen

NEW ZEALAND GRAPEFRUIT OR POORMAN ORANGE

(Sunfruit)

Of unknown parentage and almost certainly of hybrid origin, New Zealand Grapefruit has many characteristics similar to those of the grapefruit but it cannot be directly related, since it originated in China and was developed in Australasia.

It is said to have been taken by Captain Simpson from Shanghai, China, to Australia in 1820, where it was named Poorman Orange, then to Kawau Island, New Zealand, by Sir George Grey in 1855. Later propagated by J. Morrison, it was named Kawau Grapefruit or New Zealand Grapefruit, and a variant as Morrison's Seedless Grapefruit.

New Zealand Grapefruit trees are vigorous, large and prolific with much longer and narrower leaves than those of grapefruit, being similar to those of the Australian variety Smooth Flat Seville orange. In my opinion the recently developed Sunfruit in Swaziland has identical fruit characteristics to those of the New Zealand Grapefruit but the leaf shape is slightly different.

The fruit is medium-large, intermediate between a sweet orange and a grapefruit, round in shape with a slightly pebbled rind texture. The rind is moderately thick (compared with grapefruit of similar size), leathery in texture and tightly adhering, but is rarely peeled since it is normally used as a breakfast fruit.

Internal maturity may be reached before the fruit has achieved full colour – particularly in semi-tropical regions – but in New Zealand with its marginal citrus climate, full colour is reached before peak maturity. The transition from colour break to yellow is quickly achieved and rapidly progresses to orange-yellow. It reaches maturity in New Zealand in August and can be regarded therefore as being a midseason variety.

The flesh is straw-coloured, somewhat coarse in appearance but tender and extremely juicy: the juice yield is similar to that of many oranges and higher than most grapefruit. Often virtually seedless when grown in isolation from most other citrus varieties, it is moderately seedy with from six to 15 seeds when cross-pollinated.

When mature, the flavour is unique among citrus varieties, having some of the characteristics of the grapefruit combined with the slightest trace of lime and lemon, somewhat resembling a 'bitter lemon' soft drink.

Although it holds well on the tree in New Zealand, Sunfruit in Swaziland has only the shortest of harvesting periods – perhaps no more than three to four weeks – during which the fruit passes from having acceptably low acidity to being overmature and extremely susceptible to post-harvest decay.

When harvested during this period it may be successfully stored for about two months: much less than that of most grapefruit varieties.

In the 1970s a strain with deeper orange rind colour was discovered in New Zealand and named Cutlers Red. It bears close resemblance to the Japanese Hassaku and Australian Smooth Flat Seville.

It has been suggested that the New Zealand Grapefruit and the Ugli are identical, but this is not correct. Small quantities of Ugli are produced in New Zealand where they closely resemble the fruit grown in Jamaica but are more oblate and a deep orange colour. Throughout the northern part of New Zealand's North Island, the New Zealand Grapefruit is a favourite garden tree.

OROBLANCO OR SWEETIE

Oroblanco originated from the same crossing as Melogold at the University of California, Riverside.

The trees are vigorous and attain similar size to Marsh grapefruit but have broader leaves.

Fruit shape is similar to that of Marsh although slightly flatter and somewhat larger in size. However, the rind remains greener far longer and is very slightly thicker. Other internal physical characteristics are similar to Marsh but Oroblanco has somewhat paler flesh and a wider open core. The flesh is tender and very juicy but slightly less so than Marsh on account of having a thicker rind. It

OROBLANCO OR SWEETIE
The rind here is somewhat thicker than normal

is completely seedless when grown in isolation and, like its triploid sister variety Melogold, sets few seeds even when pollinated.

The flavour of Oroblanco is noticeably sweeter than Marsh due to its significantly higher sugar content – typically 2 per cent above Marsh – and acid level as much as 0.6 per cent lower. This higher sugar to acid ratio, combined with a lack of bitterness which often characterises Marsh when grown in cooler production areas, has proved a valuable asset which has been exploited by growers in Israel, where Sweetie has been planted on a significant scale in the past decade.

At present production in Israel has reached some 8,000 tons per annum and is expected to increase to 20,000 tons in the next few years. It was soon realised that quality, particularly with regard to rind thickness, juice content and flavour, was adversely affected when the tree was grown on vigorous rootstocks such as Volkameriana and rough lemon. Early plantings on such rootstocks have been removed and currently only Troyer citrange and Swingle citrumelo are employed.

While yields of Sweetie fall slightly short of Marsh, the early-maturing, sweeter-flavoured characteristics have been emphasised by the Israeli Citrus Marketing Board (CMBI) and the slow rind colour development has been made a feature of the variety. By highlighting its bright green colour and naming the variety Sweetie, consumer awareness of its good features has been created. Later in the season when the fruit is well coloured it is marketed as Golden Sweetie.

Oroblanco may be harvested four weeks or more earlier than Marsh in Israel and by several months in the Central Valley, California, just as soon as the juice becomes 'free-flowing' and achieves about 40 per cent by weight. It may not be left to hang on the tree too late in the season, because quality is adversely affected as astringency and off-flavour become more apparent.

RED SHADDOCK

The origin of this African pummelo recently developed by S. Burdette and O. Skaarup at Tambuti Estate, Swaziland, is not certain, but it has several fine features.

Of attractive appearance, the fruit is large for a pummelo, averaging about 140 mm in diameter, round in shape with a slightly pointed stem-end, and having a smooth, glossy, yellow rind without a blush. The rind is somewhat thicker than most pummelos while segment size, shape and number are typical, many being irregular or 'blind'.

However, the flesh is a deep red like that of the best pigmented grapefruit variety, Star Ruby. The texture is smooth like that of grapefruit and not slightly coarse as is sometimes found with Chandler, and is very tender and juicy. Flavour is outstanding with low acidity and a high sugar content.

RED SHADDOCK

REINKING

Originating from a cross between Kao Phuang and pollen from an unknown source, Reinking was selected by J. R. Furr at Indio, California.

The trees have denser foliage than the other commonly grown Californian pummelo variety, Chandler, and are somewhat larger in size but have similar vigour. Much of the fruit is produced inside the tree's canopy.

Reinking fruit is larger than Chandler, most in the range 600 to 1,100 g, round to somewhat elongated in shape with a very small but pronounced neck. The rind is yellow, fairly smooth in texture but uneven when compared with Chandler. Rind thickness is similar, being moderately thick and tightly adhering, with both varieties having an average of around 17 segments per fruit. Reinking flesh is white but is otherwise similar in juice content, texture, tenderness and flavour to Chandler. However, Reinking has a bitterness component which is very noticeable to some people but not to others. Both varieties are being grown on a small scale in California, particularly in the Central Valley which is well suited to the production of high quality fruit.

TAHITIAN

(Moanalua)

The origin of this variety is unclear but probably Tahiti from seed from Borneo. It is grown in Hawaii on a very limited scale.

The tree is moderately vigorous and small in size with medium to large fruit, most in the range 800 to 1,400 g and with a unique shape, being very flat with four almost completely straight faces.

Rind is very smooth and glossy, yellowish-green to yellow when mature but is fairly thick. The segments are irregular in shape and total about 18, with many being small and 'blind'.

The flesh is greenish-white, smooth in texture

like a grapefruit, tender and juicy with a distinct lime flavour and moderate acidity – not unlike a sweet New Zealand Grapefruit.

While the flavour and other internal quality characteristics are good features of this variety, it is difficult to believe it has much potential as a fresh fruit on account of its size, and the irregularity of segment shape and size.

UGLI

This unusual variety was discovered growing wild near Brown's Town in Jamaica in 1914. The name has been patented and was coined as a corruption of the word 'ugly', since this aptly describes its external appearance. Grown only in its native Jamaica, it is almost certainly a tangelo, judging by the tree's mandarin-like leaves and the fruit's size and quality.

The fruit is large – slightly larger than a grapefruit – and is pyriform in shape having a flattened and heavily depressed apical area with a noticeable and strongly furrowed stem-end. The peel is extremely rough, uneven and fairly thick. Being from the tropics, the fruit matures without the rind colour having fully developed. It is harvested from October onwards when the peel is greenish in colour. It turns yellowish-orange only towards the end of the season in April, and when stored for some weeks thereafter. A heavily blemished rind is typical.

The core of the fruit is large and usually open which, together with the thick rind, gives the fruit a light 'feel'. The segments are correspondingly small, but the orange flesh is extremely tender, has a soft texture and is very juicy.

It is customarily eaten with a spoon like a grapefruit. Seed content varies considerably; often seedless, it may sometimes have 12 or more seeds per fruit. Never acidic, flavour is sweet, delicate and orange/mandarin-like with no trace of bitterness.

It is popular in England, Canada and more recently in the Netherlands, but is very expensive, since carefully regulated supplies to the markets ensure it has a scarcity value.

When produced in a sub-tropical climate such as the Cape Province, South Africa, and New Zealand, the flavour is invariably acidic and lacking sweetness. To achieve the unique and distinctive quality, there seems no doubt production has to be limited to truly tropical growing regions.

UGLI

DJEROEK DELEEMA KOPJOR (HYBRID 202) PUMMELO. IT COMBINES OUTSTANDING QUALITY AND GOOD PRODUCTIVITY

THE LEMON

Citrus limon

Recently it has been suggested that the Mediterranean or Italian lemon we know today is a hybrid of the citron, the Indian lime and a third as yet unknown citrus type (probably pummelo). Although it has never been found growing wild in the foothills of the Punjab, many other natural hybrids such as the Gal Gal or Hill lemon and the rough lemon are indigenous to that region.

The lemon appears to have originated in the Punjab region of Pakistan and India from one such hybrid Indian lemon, with further development having occurred possibly in one of the citrus gardens in Media (now Iran) or elsewhere in the Middle East. Lemons were unknown to the Romans and were first taken to the Mediterranean area – Spain, in fact – by Arabs around AD 1150.

Early lemon production in the Mediterranean started near Genoa on the Ligurian coast and as the climate there is far too cold without protection, plantings spread south to the Amalfi region near Naples and finally to the southern tip of mainland Italy and to Sicily. From there the lemon spread throughout the Mediterranean, then to the New World with Columbus on his second voyage in 1493, and eventually worldwide.

Although lemons will grow well under tropical and semi-tropical climatic conditions, world production is largely restricted to sub-tropical regions since the tree and fruit present problems in the more humid environments due to pests and diseases. Moreover, the small acid lime *Citrus aurantifolia* is regarded as being a perfectly adequate substitute and is far better adapted to the tropical conditions.

FRUITING BRANCH OF LAPITHKIOTIKI VARIETY IN CYPRUS

Lemon trees grow vigorously and become large if not pruned. They are more cold-sensitive than oranges, grapefruit or mandarins. Moreover, unlike these other citrus species which normally bloom just once a year, lemons will produce a series of flowering flushes throughout the season.

While some varieties are more everbearing than others, it is common to find fruit at different stages of growth on the tree at the same time. This repeated-flowering, everbearing characteristic has been exploited in some parts of the world, particularly in Italy and to a lesser extent in Spain, where the tree growth is manipulated by cultural practices in such a way that larger than normal crops are produced when fruit is scarcest. Perhaps the best example of this is the Verdelli lemon (or green summer lemon) crop in Italy.

The Verdelli crop is normally set in August, but is invariably light representing no more than about 10 per cent of annual production. However, if irrigation to the trees is withheld during June and the tree severely wilted, heavy flowering can be induced and a much larger Verdelli crop set when irrigation and fertiliser are applied following the wilting period. The practice, known as Forzatura, has been adopted by Sicilian lemon growers for well over a century. The Verdelli crop remains on the trees for nearly 12 months and is picked while still green the following summer.

VERDELLI

In addition to a flowering rhythm which can be manipulated in this way, the lemon fruit is amenable to storage for long periods. Using sophisticated storage techniques, much of the Californian lemon crop is harvested before full colour has developed and is then successfully held under controlled temperature and humidity conditions for between several weeks and several months to ensure a more even distribution of fresh fruit supplies.

A less sophisticated method is used in Turkey with Yatak lemons which are stored in naturally ventilated caves from November and December until the following summer, although it has to be admitted that high losses are encountered and the condition of sound fruit is rather precarious.

On a global basis, lemon production stands at around 4.9 million tons. The United States leads with 990,000 tons, followed by Italy with 880,000 tons and Spain with 650,000 tons. Argentina leads Southern Hemisphere production with 400,000 tons. About 60 per cent of the United States' production is processed, while in Italy and Spain the proportions are 22 per cent and 10 per cent respectively.

While many different varieties of lemon have been established and show considerable variation in horticultural characteristics, there are not the same differences in fruit traits as are found within oranges and mandarins, for example. It is often almost impossible to identify with certainty fruit of a particular variety, since the range of different fruit shapes is often as wide from different flushes of the same tree as it is between varieties at the same time of year.

The leading lemon varieties are as follows.

BEARSS

(Sicilian)

The Bearss or Sicilian variety, a recent selection (1952) from an old tree at Bearss Grove, near Lutz, Florida, is harvested between July and December from trees which are very vigorous, thorny and very sensitive to cold.

The fruit is susceptible to peel injury when harvested early in the season and rapidly breaks down internally following freezing weather. It typically has up to six seeds per fruit.

Few lemons are produced in Florida where extra large size and keeping quality present problems. However, Bearss is the only variety recommended for the climatic conditions there, and where much of the crop is processed for its rind oil as well as for its juice.

EUREKA

EUREKA

Outside the Mediterranean basin Eureka is the most widely grown lemon variety, forming a major part of the crop in California, Australia, South Africa and a significant proportion of Argentina's production. Most of Israel's lemons are the Eureka variety also.

Despite some shortcomings it has many favourable characteristics which have led to its being planted on such a wide scale. Of Californian origin from seeds imported from Sicily in 1858, it was first released as Garey's Eureka but soon became known simply as Eureka.

The tree has a spreading habit but is only moderately vigorous and is much smaller than the other widely grown variety, Lisbon. Only sparsely foliaged and markedly less cold-resistant than Lisbon, it is well adapted to the coastal areas in Australia and California where frost damage is more rarely experienced. Being smaller, it is often less productive than Lisbon but has a well-distributed harvest season through late winter, spring and early summer. There is a marked tendency to produce the fruit in terminal clusters, with the result it is often more prone to wind blemish and sunburn than the Lisbon or Fino varieties, where the fruit is well protected inside the canopy of the tree. Eureka trees are less thorny than Lisbons, making for easier picking than most other lemon varieties.

Eureka is incompatible with trifoliate orange rootstock and citranges (except the Australian Benton).

Eureka fruit is somewhat smaller than the other important lemon varieties Lisbon and Verna. Although the rind is smooth when grown in many locations, in Spain and in other areas with similar climatic conditions it may be coarser and the main crop deteriorates in quality and becomes puffy two or three months after reaching maturity. However, it is well suited to the 'curing' process in California and keeps well in cold store.

Rind thickness is medium to thin, and the fruit has a high juice content with a high acid level. Seeds are few, rarely more than five, and often the fruit is seedless. Although outyielded in Spain by both the Lisbon and Fino when fully grown, Eureka production is superior to both in the early years.

Although its popularity has waned in California in recent years in favour of Lisbon, Eureka has been planted on an increasing scale in Spain where, until the mid-1970s, it formed less than 5 per cent of production. It is now budded increasingly on macrophylla rootstock in California and Spain, although not without some disadvantages.

Several selections have been developed, the most common being Frost. However, in recent years almost all the Eureka lemon plantings have been of the Allen strain.

FEMMINELLO COMUNE

(Femminello Ovale)

The Femminello comprises a group of several selections each of which has its own particular characteristics. They all have good vigour and productivity as well as being everblooming and everbearing.

Collectively Femminello selections account for about 75 per cent of the Italian crop. All Femminello varieties produce four crops per year which are each given distinct names so that in the fresh fruit trade the flowering or crop is referred to rather than the particular variety or selection from which it originated. The crops from the various flowerings are as follows:

	Crop	*Period of harvest*
1	Primofiore	September–November
2	Limoni	December–May
3	Bianchetti	April–June
4	Verdelli	June–September

The distribution of the crop over the four different harvests is determined by the climatic conditions and other factors as well as the characteristic of each of the Femminello selections. For example, Femminello Comune would typically have 76/12/12 per cent distribution over the Primofiore + Limoni, Bianchetti and Verdelli crops respectively, the Femminello Siracusano 85/5/10 per cent and the Femminello St Teresa 60/10/30 per cent.

Besides being productive, Femminello selections produce fruit of medium size, moderately thick peel, and of correspondingly lower juice content but with higher acidity than other varieties. Seed content varies depending upon the crop: Primofiore and Limoni having typically five to 12, Bianchetti two to ten and Verdelli three to eight seeds per fruit. While the juice percentages of Limoni and Bianchetti are similar, the Verdelli crop is usually significantly lower due mainly to the juice not being free-flowing or easily extractable.

FEMMINELLO COMUNE

Almost all Femminello selections are very susceptible to mal secco disease, which is widespread throughout most Italian lemon groves, but the selection Femminello St Teresa shows some tolerance and has been planted on an increased scale in the past two decades. It is believed the variety in Turkey known as Italian is probably Femminello St Teresa.

GENOVA

(Genoa)

Like the Villafranca, this variety is also of Italian origin being exported first to California then to Florida about 1881. The tree is thornless and of smaller habit than Eureka, but is more cold-resistant and has denser foliage.

The fruit is just as smooth but is more spherical with a small pointed neck and nipple. Internally the fruit has similar quality to Eureka: juicy, acidic, thin rind and variable seed content ranging from seedless to six seeds per fruit. In California, Genoa is considered to be a strain of Eureka.

It is grown principally in South America: in Chile, for example, it is the leading lemon variety and also forms a significant proportion of the Argentinian crop.

INTERDONATO

(Speciale)

Originating on a Colonel Interdonato's property in Nizza, Sicily, in 1875, this is the earliest maturing of lemon varieties, being at least four weeks ahead of the Femminello.

Like the Monachello, citron parentage is almost certainly involved in the Interdonato genetic make-up, and it is also resistant to mal secco. The tree is moderately vigorous and nearly thornless, with the leaves resembling to some extent those of the citron.

This distinctive variety forms about 5 per cent of Italian production but a significantly higher percentage of Turkey's 250,000 ton lemon crop. It almost certainly has the lowest juice content of any major lemon variety due to some extent to the widespread practice of harvesting it well ahead of maturity. The fruit is much larger, longer and smoother than other lemon varieties, with a thin peel and few seeds. Acidity of the juice is similar to Monachello.

Unlike the true lemons which have from nine to 11 segments, the Interdonato often has only six or seven. The rind is thinner than most Mediterranean lemons but the fruit has a low juice content.

Rind colour is improved by degreening but the nipple, which is a susceptible area on this variety, is invariably turned brown in the process. In late December the Interdonato is overmature and is inclined to drop from the tree.

ITALIAN

This Turkish variety so closely resembles the Femminello St Teresa in both fruit shape and tree tolerance to mal secco that they are believed by most authorities to be the same selection.

Grown extensively in the Mersin district, the fruit has a roundish shape, moderately thick peel, and is very seedy. It also has particularly firm condition quite unlike most other lemon varieties. It is a reliable bearer and the juice has a high acid content.

GENOVA

INTERDONATO

ITALIAN
This is probably the variety Femminello St Teresa

LAMAS

LAMAS

Grown on a significant scale in Turkey, this variety bears just one crop each year, harvested during the November to December period. Fruit characteristics are similar to Femminello Comune in shape, peel thickness and seediness.

Its productivity is less reliable than the Italian variety, tending to alternate-bearing, but the fruit is particularly suited to the 'Yatak' storage programme (Yatak in Turkish meaning bed), in which lemons are stored in basalt caves in the Urgup and Goreme regions of central Turkey.

LAPITHKIOTIKI

Of unknown origin, Lapithkiotiki is the principal lemon variety planted throughout Cyprus, deriving its name from the village of Lapithos near Kyrenia on the north coast.

Lapithkiotiki trees are vigorous with a spreading habit, necessitating their being more widely spaced than other lemon varieties. They are very productive, fairly thorny and reasonably tolerant to mal secco disease, especially when compared to Eureka and Lisbon but less so than St Teresa. In spring Lapithkiotiki trees shed a higher proportion of their leaves than other lemon varieties.

In Cyprus the trees essentially set one crop each year which matures from mid-September onwards. During the early part of the season Lapithkiotiki fruit may be degreened without the problems often encountered with Eureka.

The fruit is not unlike Eureka in size and quality, having similar rind texture and thickness, juice content and freedom from excessive seediness. Its shape, however, is somewhat more tapering at both ends. Lapithkiotiki is harvested throughout the six-month period ending in late March without appreciable loss of quality or condition. From April onwards the fruit becomes overmature, turning bronzy-yellow in colour and developing a hollow core, although juice content is little affected, but post-harvest decay becomes increasingly problematical. If left unpicked until late June, much of the fruit drops to the ground.

LISBON

Lisbon is of Australian origin from seed exported from Portugal before 1824, but its popularity in California followed introductions in the 1870s.

Of increasing popularity in California and widely planted there and in Arizona desert areas, it is also of importance in Australia, Uruguay and Argentina. Lisbon is more cold-resistant than Eureka as well as being more productive, more vigorous and with denser foliage. However, the trees are thornier. Fruit is produced on the inside of the tree where it is protected from sun, wind and cold damage. Lisbon will often outyield Eureka by as much as 25 per cent, partly due to its large size and bearing area, but it has to be more widely spaced. Production is restricted mainly to winter and early spring.

Unlike Eureka, Lisbon is compatible with trifoliate orange and citrange rootstocks.

In comparison with Eureka, differences in tree characteristics are far greater than those in the fruit, although the Lisbon has a somewhat less pronounced nipple and slightly rougher rind texture. Juice percentage and acid level of Lisbon are similar to those of Eureka. When marketed from California no distinction is made between these two varieties.

There have been several selections made in California such as Prior, Rosenberg and Strong, but the most successful ones are Limoneira 8A and Frost.

LISBON
Currently the most popular lemon variety in California

MEYER LEMON

There is little doubt that this variety is not a true lemon and is probably a hybrid between the lemon and either an orange or a mandarin. First imported into the USA from China in 1908 by F. N. Meyer, the tree is grown almost exclusively as a 'dooryard' plant and is popular for its small compact size, its cold-resistance which is far better than any true lemon, and its production of fruit year-round (although mostly during the winter and early spring period).

MEYER LEMON

The fruit more closely resembles an orange than

a lemon internally as well as externally. It is fairly large – 65 to 75 mm in diameter – almost round in shape and with only the smallest of nipples, usually almost inconspicuous. It has an attractive appearance, with yellowish-orange rind which is exceptionally smooth, soft and thin but lacks the typical lemon peel oil aroma; in fact the variety is not desirable for lemon oil production.

The pulp is usually a dark yellow colour, very juicy and tender, containing about ten small round seeds per fruit. The flavour is distinct, as if having the sweet lime in its parentage, and it has a lower acid level than the true lemons. This variety is very sensitive to handling damage probably because of its smooth rind texture and soft condition.

MONACHELLO

Many of this Italian variety's fruit and tree characteristics suggest it is a lemon–citron hybrid. The tree's lack of vigour and distinctive round-topped shape suggest part citron parentage, while the thick peel and low juice percentage are further indications. The juice has a lower acid level than most other true lemons, although it is still quite acceptable for most culinary purposes.

The outstanding characteristic of the Monachello is its resistance to mal secco disease and it has been planted in areas of Italy where the disease is most severe. Currently Monachello accounts for about 10 per cent of Italy's lemon production. It is not grown outside Italy.

PONDEROSA

(American Wonder)

Of very limited use and no commercial importance, the Ponderosa is almost certainly a hybrid of the lemon and citron with both the tree and fruit closely resembling the latter. The tree's small habit makes it a popular garden variety in the USA; in California the Ponderosa is often grown as a garden ornamental espaliered against a wall, for which purpose it is well adapted. The exceptionally large size of the fruit, the rind of which has a rough texture and is fairly thick, could easily lead it to be mistaken for a small pummelo.

The flesh is moderately juicy, fairly acidic and contains many seeds. Fruits mature throughout the year and are often used as a substitute for the true lemon. Being a citron hybrid it is more cold-sensitive than other lemons but not as tender as the limes. The trees are very thorny, everflowering and everbearing.

MONACHELLO

PONDEROSA

PRIMOFIORI OR FINO

(Mesero, Blanco)

Primofiori is of unknown origin but possibly came from the old Spanish variety Comun de la Vega Alta del rio Segura. It should not be confused with lemons of a similar name (Primofiore) from Italy, which refers to fruit produced by all varieties from the main or first (or spring) flowering and which have a similar harvest period as this Spanish variety.

Primofiori trees are vigorous, thorny and highly productive, yielding uniformly each season, reaching maximum production at around 14 years, and may continue to 50 years or older.

Primofiori has regular-shaped fruit, being spherical to oval, with a smooth, thin rind and a relatively short nipple when compared with Verna. Fruit is smaller but juice percentage higher than the Verna except at the very beginning of the season when the fruit is harvested before full maturity. At this time it is usual to degreen the fruit to obtain acceptable colour. Primofiori has more seeds than Verna, averaging about five per fruit, and the acid level is also somewhat higher, averaging about 7 per cent.

Because it is the first maturing variety in Spain it is grown mainly in the valleys where frosts are most likely to occur. It accounts for around 20 per cent of Spanish production and like the principal variety, Verna, is often selectively picked to size on two or more occasions, the small fruit being left on the trees irrespective of colour to increase in size. However, unlike the Verna, the Primofiori produces no secondary crop.

Primofiori, like Verna and Villafranca, has proved highly susceptible to mal secco disease when introduced into Italy and other regions of the Mediterranean where this disorder is prevalent.

PRIMOFIORI OR FINO

VERNA

(Berna)

Verna is a variety of unknown origin and somewhat surprisingly has not been planted to any extent outside its native Spain. The Verna variety presently forms around 70 per cent of Spain's increasingly large lemon crop of around 645,000 tons, most of which is produced in the Murcia district, Alicante Province.

Verna trees are large, spreading and virtually thornless. They usually flower twice but in some years a third crop is produced. The second crop – the Secundus – is of inferior quality and of little commercial importance. Growers will sometimes force the tree to produce a larger third crop, known as Verdelli or Rodrejos.

The main crop or Cosecha comprises medium to large fruit with a pronounced nipple and usually a well-developed neck. The rind is medium thick to thick, and somewhat rough and uneven in texture. The juice yield is correspondingly lower than other varieties, but the pulp is tender and the juice of good acidity. There are few seeds. The Cosecha crop extends from late February until late July or early August, and the variety is able to hang on the tree in good condition without marked deterioration in quality under Spanish conditions.

The Verdelli crop from Verna trees has fruit which is rounder with a much thinner and smoother rind, and as its name suggests often a significant portion of the rind remaining green at harvest. However, it is often degreened before being packed and marketed. The Verdelli

VERNA
Spain's most important lemon variety

(Rodrejos) crop is harvested from the end of August to October.

Verna trees tend towards alternate-bearing, particularly following the 'Verdelli treatment', and are outyielded by the Primofiori (Fino) variety. Despite these shortcomings, it remains an outstanding variety with its ability to produce fruit throughout much of the summer when lemons are in shortest supply on European markets.

VILLAFRANCA

Introduced into Florida from Sicily in 1875, the Villafranca, until recently, constituted most of the lemon crop in that State. It has now been superseded by the variety Bearss (or Sicilian).

Villafranca is of minor importance in most producing areas, although at one time it was the leading variety in Queensland where it produces a good summer crop. It is still an important variety in Israel where it produces more seedy fruit than the leading variety Eureka.

The tree is more vigorous than Eureka and is thornier while young but is almost impossible to differentiate the two varieties by mature tree habit and fruit characteristics.

THE LIME

Limes comprise a varied group of types which are so different from one another in tree and fruit characteristics that they have been given separate species status. It is convenient to divide them into two broad groups: the acid or sour limes of which there are two important types, and the sweet or acidless limes.

Of the two acid limes, the smaller-fruited, very seedy one (*Citrus aurantifolia*) is known by several names: West Indian, Mexican and Key are the three most widely used. It is grown only under extremely hot climatic conditions free from frost hazard. Although produced on an extensive scale in Mexico, Brazil, Egypt and India as well as in other countries worldwide, the West Indian lime is virtually unknown in many importing countries, particularly throughout Europe. Far more familiar there but nevertheless imported in only limited quantities is the larger-fruited seedless Persian or Tahiti lime, *Citrus latifolia*, which is produced mainly in Brazil and Florida.

The sweet lime (*Citrus limettioides*), commonly referred to as the Indian or Palestine sweet lime, is grown on a limited scale over a wide area including the Middle East, the Indian subcontinent and some parts of Central and South America. This lime should not be confused with what is sometimes called the Mediterranean sweet lime which is, in fact, the Tunisian or Mediterranean sweet limetta (*Citrus limetta*).

FRUIT AND FOLIAGE OF WEST INDIAN LIME

ACID OR SOUR LIMES

PERSIAN, TAHITI OR BEARSS LIME

(Citrus latifolia)

Persian limes are thought possibly to have been introduced to the USA through San Francisco from Tahiti some time between 1850 and 1880. Despite its name this variety is not grown now in either Tahiti nor for that matter in what was formerly Persia (Iran).

Although believed to be a hybrid between the small acid lime and possibly the citron, its true origin is obscure, having probably come from the Orient by way of Persia, the Mediterranean, then possibly via Brazil and Australia (where it was reported to have been grown as early as 1824), thence to Tahiti and finally to California in the second half of the 19th century. To complicate this history further, the Persian lime is not known by that name in California but is referred to as the Bearss lime. The Bearss, Persian and Tahiti limes are the same variety.

Persian lime trees are larger than the West Indian type, are nearly thornless and have larger, darker green leaves. They are also far more resistant to cold, but more sensitive than the lemon.

In Florida, Persian lime is often propagated as marcots or air layers because they come into bearing at an earlier age than budded trees.

Fruits of the Persian lime, particularly the smaller ones, may often be mistaken for those of West Indian lime since externally they are virtually indistinguishable; moreover, both may sometimes be mistaken for small immature lemons.

It was for this latter reason that Persian limes were and still are picked and marketed when dark green in colour since the fully mature fruit is greenish-yellow or pale yellow, like the West Indian lime. The rind is very thin with the distinctive lime rind oil aroma when the surface is

PERSIAN LIME (LEFT) WEST INDIAN LIME (RIGHT)
The Persian lime is seedless and larger than the West Indian variety

scratched. Persian limes are susceptible to citrus scab. If left on the tree until past peak maturity the fruit is yellow and unsaleable, and tends to develop a rind breakdown problem at the stylar-end.

Since the Persian lime is a triploid and produces no viable pollen, the fruit is almost always completely seedless. The greenish-yellow flesh is very juicy and extremely acid: levels are often similar to the West Indian lime. It is the only lime produced on any commercial scale in the USA; almost all are grown in Dade County in south-east Florida, where freezes are rarely experienced. In California the Persian (Bearss) lime is produced in the south of the State but on a smaller scale.

Fresh lime exports to Europe originate primarily in Florida and Brazil, although small trial consignments are presently being sent from Israel: they are all of the Persian variety. In Australia production is increasing and is limited to the eastern coastal areas since inland areas have a marked frost hazard. Much of the fruit is used fresh and is regarded as being unsuitable for lime cordial products.

West Indian, Mexican or Key Lime

(Kagzi nimboo, Limun beladi, Limão galego, Lima)

(Citrus aurantifolia)

This variety is often referred to as the 'true lime', as well as by the three names above. Originating in the Malaysian region of south-west Asia, it has been cultivated in many countries worldwide where climatic conditions permit its growth.

The tree is fairly vigorous and of medium size and somewhat bushy growth habit, with slender branches covered with many small thorns. The leaves are small, pale green in colour and blunt-pointed, while the tree as a whole is extremely sensitive to cold: far more so than the Persian lime.

Unlike most other citrus species, West Indian lime is usually propagated from seed, particularly in India, Egypt and Mexico. Some different types

West Indian Limes

These turn yellow when fully mature but are often harvested when they are still green

have been selected, but most (which form the majority of world production) are indistinguishable.

The fruit is small, roundish in shape with a small nipple and a very slight neck. The rind has a smooth texture and is extremely thin, and the rind oil has the distinct pungent aroma which typifies both the West Indian and Persian limes.

When the fruit is allowed to mature on the tree the rind is greenish-yellow or yellow, but it is often harvested earlier while still dark green in colour. The flesh is a light greenish-yellow, tender when fully mature and very juicy, but sometimes less so when harvested earlier. It is highly acid, having as high, and sometimes higher, citric acid levels as lemons. Whereas Persian limes are seedless, the West Indian variety can sometimes be very seedy, often with 15 or more seeds per fruit.

In tropical areas, West Indian lime will flower throughout the year depending upon rainfall distribution but in the sub-tropics it bears one crop which usually matures in late summer and is overmature by mid-winter. When West Indian limes are allowed to become overmature on the tree, the fruit soon drops on the ground thereafter.

Mexico leads world production, with some 600,000 tons per annum, most of which are grown in the west coast states of Colima, Michoacan, Guerrero and Oaxaca and to a far lesser extent in Tamaulipas on the Gulf of Mexico. The lime trees are often in mixed plantations with coconut palms, mangoes or papayas. Some of Mexico's tonnage is processed, but more than 85 per cent is sold on the local fresh market and some exported to the USA.

In Egypt, the third largest producer of West Indian limes, a forcing practice is employed similar to that used to increase the summer or Verdelli lemon crop in Sicily (see page 88). Most limes there are grown under irrigation south of Cairo, near El Faiyun, and sold fresh locally.

Much of the West Indies' production is processed, most by a method which results in a distinctive form of lime juice greatly appreciated by the British. The whole fruit is crushed with rollers and, after screening, the juice is held for periods up to 30 days. Normal fermentation is inhibited by the high acidity, low sugar content and the presence of peel oil. The juice is later drawn off from the residue. It has a distinct musty aroma quite different from that of freshly-squeezed lime juice or processed juice sold in the USA and Canada.

Although exported to the USA, where it is called the Key or 'Bartender's' lime, there is no commercial production of this variety in Florida or elsewhere throughout the country, since only the Persian or Bearss lime is grown in the USA.

In Brazil, where it is known as the Galego lime, annual production totals approximately 300,000 tons, most of which is sold fresh on the local market with less than 1 per cent being processed. They are not exported fresh on any scale; lime exports to Europe are of the Persian type.

In Iran (formerly Persia) only the West Indian variety is grown, and is known locally as Shirazi or Torsh (meaning acid); it is grown mainly near Fars in the Shiraz region and in the south in Bandar Abbas.

OTHER SMALL-FRUITED ACID CITRUS VARIETIES

Although not related to limes, this group of small-fruited acid types are of importance in the Far East. It includes Yuzu, grown in China and Japan, and its hybrids Kabosu and Sudachi which are uniquely Japanese citrus fruits.

KABOSU

(Citrus sphaerocarpa)

Grown principally in Oita and Tokushima Prefectures, Kabosu fruit is slightly larger than Yuzu and quite different in appearance both externally and internally. It is almost round, with a thin, smooth, finely pebbled rind. Most fruit is harvested during August and September while the rind is dark green. It may be successfully stored in polyethylene-lined boxes without loss of colour or condition for very long periods. Kabosu is sold out of store throughout the winter until April.

The flesh is juicy and, like Yuzu and Sudachi, is very acidic. It is fairly seedy, averaging around 12 seeds per fruit. Kabosu is often served in cut halves as a garnish.

Produced on about 700 ha, the annual crop totals nearly 5,000 tons.

MAKRUT LIME OR COMBAVAS

(Citrus hystrix)

The Makrut lime is widely grown throughout south-east Asia from Sri Lanka to the Philippines. It is usually produced on a small scale with the fruit used locally for flavouring food and other purposes.

The tree is small with many thin branches. When fully mature the fruit is yellow but almost invariably it is harvested immature while still green. The Makrut lime is small, with an unusual and distinct shape and rind texture. The stem-end is pronounced and the rind of irregular and extremely bumpy texture; Malaysians liken its appearance to that of a crocodile's eyebrows!

The rind is thick and its oil has a very pungent lime-like aroma when picked immature. The fruit is almost completely devoid of juice which is both bitter and very acidic. Makrut limes are usually very seedy.

Used primarily for its rind oil, the Makrut lime – and its leaves – enhances the flavour of curries, soups and salads. Its other uses are in the preparation of a hair rinse which is faintly perfumed and reportedly has insecticidal properties. It is also used for smearing on the feet to kill land leeches.

Occasionally small consignments are exported by air to Paris where they are called Combavas: probably a corruption from a closely related species, *Citrus combaras*.

SUDACHI

(Citrus sudachi)

Many features of the Sudachi are similar to the Kabosu but the size is much smaller, averaging between 30 and 40 g compared with 100 to 150 g. Both types are grown in the same regions, har-

MAKRUT LIME OR COMBAVAS

vested in late summer and stored in the same manner. They are marketed over the same August to March period. Sudachi is also used in much the same way as Kabosu and is grown on a similar scale.

YUZU

(Citrus junos)

In Japan, Yuzu is grown for both its fruit and as a rootstock for other citrus varieties. Its characteristics as a rootstock are described on page 123.

Yuzu originated in China and was introduced to Japan possibly more than 1,000 years ago. More cold-hardy even than the satsuma mandarin, Yuzu has a wider distribution in the colder areas of Japan.

The fruit is small, averaging about 100 g, but is lightweight for its size. Rounded in shape, it has an uneven, rough textured yellow rind which is quite thick. Although not very juicy, it is the rind which is most often used in many Japanese dishes. The common Yuzu is extremely seedy but a somewhat smaller seedless selection is now widely grown for the fresh fruit trade. Yuzu is harvested from October to March and, like Kabosu and Sudachi, is also used to make vinegar and for flavouring soup.

YUZU
The Yuzu shown here is overmature internally

There are around 1,800 ha in commercial Yuzu production, mainly in Kochi and Tokushima Prefectures, with an annual crop of 9,000 tons. Yuzu is also commonly planted as a backyard tree throughout the citrus-growing regions of Japan.

SWEET LIMES

(Citrus limettioides)

PALESTINE OR INDIAN SWEET LIME

(Mitha nimboo, Limun helou)

Like other 'sweet' citrus fruits such as acidless or sugar oranges, citrons and sweet limettas, sweet lime has in fact a somewhat lower sugar content than acid varieties, but because it is almost completely devoid of acidity it is referred to as 'sweet'. In reality it is the most insipid of citrus fruits. While West Indian and Tahiti limes might have 6 per cent citric acid content and oranges around 1 per cent, Palestine sweet lime frequently has less than 0.1 per cent.

The Palestine lime tree has a distinctive growth habit; it is medium sized but the branches are irregular, thick and thorny, while the leaves are cupped or rolled.

The fruit is medium in size and roundish in shape, often with a small nipple. The rind is very smooth, sometimes faintly ribbed corresponding with segments, and light yellow changing to orange-yellow when fully mature. It is tightly adhering but peelable and the rind oil has a distinctive aroma. There are only a few seeds in each fruit.

The Palestine sweet lime is very juicy and tender but, like acidless oranges and sweet limettas, it does not appeal to everyone's palate, and for this reason only a limited quantity enters into international trade. Although no reliable statistics are available on world production, the Palestine sweet lime is very popular in the Indian subcontinent and throughout the Near and Middle East. In India it is known as Mitha nimboo, and in Egypt as Limun helou or Limun succari. It is particularly favoured by children and by those who believe it has special medicinal properties.

PALESTINE OR INDIAN SWEET LIME

The Palestine lime has been widely used as a citrus rootstock in India and in Israel where it was found in the past to be particularly suitable for the Shamouti orange.

IMMATURE FRUIT OF THE CITRON

The derivation of the name citron (and citrus) is uncertain. The *Oxford English Dictionary* cites the French *citron*, the Italian *citrone* or *cedrone* (possible augmentatives of the Latin *citrus*, meaning citron tree). The word also exists in Ancient Greek: kitron (κιτρον).

Some lexicographers believe that the Latin word, and/or the Ancient Greek one, is a corruption of the Ancient Greek kedris (κεδριϲ), meaning cedar cone. This would give credence to a popular theory that the citron gained its name originally from the 'Cedar of Lebanon' (*Cedrus libani*) because of the resemblance of its immature fruit to the cedar cone in shape and colour.

Another possible explanation is that the word citrus was applied to the citron tree because the fragrance of its fruit resembled that of the wood of the North African conifer known to the Romans as citrus and highly prized by them for furniture-making. Almost certainly this was *Callitris quadrivalvis* now known as *Tetraclinis articulata*, a member of the family *Cupressaceae*.

THE CITRON

Citrus medica

Rarely grown today and unfamiliar even to many citriculturists, the citron has for several reasons a special place in the history of citrus. Citrons are indigenous to the foothills of the Himalayas and were carried first through the Middle East by the Medians during the seventh century BC.

The distinguished Swedish naturalist Carl Linnaeus gave the citron the scientific name *Citrus medica*, Media being part of what is now Iran and the region where the Greeks first became familiar with citrus fruit around 320 BC. In all probability, the citron was the only citrus fruit with which the Greeks and Romans were acquainted. Both appreciated the distinctive and pervasive aroma of the citron, employed the fruit as a perfumant, and the leaves as well as the fruit as a moth repellant!

The citron is known by the following names:

Cédrat	French
Cidra	Spanish
Cedro, Cedrone	Italian
Bushukan	Japanese
Cheu Yuan	Chinese

Citron trees have an irregular shape and attain only small size compared with lemons, oranges and grapefruit. They have a distinctive leaf shape which is long, oval, slightly rumpled and with slightly serrated margins. The trees are highly sensitive to frost injury and are often protected during periods of extreme cold by straw matting covers.

Variable in shape and usually large in size, citron fruits often have a persistent style. The rind is yellow, with a smooth but very uneven surface, and is extremely thick and tightly adhering.

Citrons cannot be peeled by hand in the same way as other citrus fruits. The flesh is lacking in juice, which may be either acidic or sweet.

Although citrons have been used in rituals since biblical times, their principal use is in the production of candied peel. The fruit is halved, the pulp removed and the rinds immersed in brine (traditionally sea water) for about a month to ferment. They are then washed and shipped in brine to Europe and North America for candying.

The citron is also of importance in the parentage of the lemon as described on page 87.

Present-day world citron production is estimated to be no more than a few hundred tons, the principal varieties being:

BUDDHA'S HAND CITRON

(Fingered Citron)

As the name implies, the fruit is split down its length into a number of finger-like sections, resembling a human hand. Much esteemed in China and Japan, where it is used for perfuming rooms and clothing, it has long been used as an adornment at religious ceremonies. Although the tree of Buddha's Hand is small and not particularly attractive, it is a popular garden shrub.

CORSICAN

(Cédrat de Corse)

Referred to as a sweet citron because of its lack of acidity, the Corsican is of similar size to the Diamante. The Corsican has a much rougher and bumpier texture and a less pronounced nipple than the Diamante, although the rind is ridged to the same degree. Other rind characteristics such as colour, thickness and texture of albedo are similar to Diamante.

The Corsican citron is also used in Bastia, Corsica, to flavour a kind of liqueur called Cédratine.

DIAMANTE

(Cedro di Diamante)

This acid-flavoured citron is the principal variety grown in Calabria, Italy, and develops into a small tree which is open in habit, has large leaves and long thorns.

Diamante fruit is typically broad and flattened at the stem-end and has a long tapering nipple. The peel is extremely thick (over 70 per cent by weight of the whole fruit), lemon yellow in colour, and smooth but slightly ridged. The albedo is firm and fleshy. Like the Corsican, it is exceedingly seedy, with 60 or more seeds per fruit. The flesh is coarse and lacks juice which is as acidic as that of the lemon.

BUDDHA'S HAND CITRON

ETROG

ETROG

(Ethrog)

Etrog trees are smaller, less vigorous and productive than other citron varieties and the leaves are more rounded and cupped.

The fruit of the Etrog is medium-small in comparison with other citron varieties, but nevertheless can still attain quite a large size. Roughly lemon-shaped, it is strongly ridged, but otherwise the rind is reasonably smooth and normally very firm. When fully mature the colour is yellow or yellowish-orange, and the peel oil has a distinctive aroma. The rind is extremely thick but cannot be removed by peeling.

The segments are correspondingly small and lack juice, which has a strongly acid flavour like that of the lemon. Seeds are excessive in number, small and brownish in colour.

For religious rituals, e.g. the Jewish Feast of Tabernacles, the Etrog citron fruit must not only be fresh and of unmarked appearance, but it must have a persistent style and be borne on a tree raised from seed or a cutting.

NAGAMI OR OVAL KUMQUAT IN ISRAEL,
SHOWING MATURE AND IMMATURE FRUITS

THE KUMQUAT AND LIMEQUAT

The Kumquat or Cumquat, known as Kinkan in Japan and Chuikan in China, was unknown to Europeans until 1846 when the English plant hunter Robert Fortune collected samples for the Royal Botanical Gardens. Although closely related to other citrus species, it has been classified by taxonomists as a separate genus, *Fortunella* (after Fortune).

Three varieties are grown for their fruits – Oval, Round and Large Round – while a fourth, the Hong Kong or Golden Bean (*Fortunella hindsii*), has extremely small pea-size fruit and is usually grown only as an ornamental shrub.

Kumquat trees have exceptional cold-hardiness due at least in part to their long dormancy extending from late autumn until early spring. However, unlike the trees, the fruits are more susceptible to frost than most other citrus fruit, a condition perhaps aggravated because of their small size.

Popular in both China and Japan, the fruit is less well known in Western countries although more and more retailers are offering them for sale. Most supplies to European markets are air-freighted from Morocco, Israel, Florida and Brazil. Smaller quantities are also sent from Corsica and the French Riviera.

Limequats result from crosses between the sour Mexican lime and kumquats. There are three named varieties. Eustis and Lakeland have the Round kumquat as one parent and are the most popular varieties. The third, Tavares, has Oval kumquat parentage but is rarely grown. Usually grown as ornamental shrubs, the fruits of Eustis have recently been exported from Israel to Europe while still green to be used in the same way as limes.

KUMQUATS

NAGAMI OR OVAL KUMQUAT

(Luofu)

(Fortunella margarita)

This is the kumquat most commonly available in Europe and North America and is grown mainly in Morocco, Israel, Brazil, California and Florida.

Of small, bushy appearance, the tree grows to about 3 to 4 m in height, although specimen trees can eventually attain great size. It has fine stems, few thorns, and smallish, dark green, pointed leaves. It is slightly more cold-hardy than Meiwa variety but somewhat less so than Marumi.

The fruit is oval in shape, slightly wider at the stylar-end, the size varying from around 20 to 30 mm long and about two-thirds as wide, and weight ranges from approximately 5 to 20 g each.

The rind is smooth, well coloured and the rind oil has a slightly more pronounced flavour than other varieties. It usually has just four or five segments and from two to five seeds. The juice, like the other varieties, is acidic, but its combination with the spicy outer rind and sweeter albedo gives a very pleasant flavour.

NAGAMI OR OVAL KUMQUAT

MARUMI OR ROUND KUMQUAT

(Luowen)

(Fortunella japonica)

MARUMI OR ROUND KUMQUAT

Marumi kumquat is grown to a far lesser extent in the Mediterranean basin and in the USA than Nagami. The tree characteristics are similar to Nagami except that it is somewhat thornier, has slightly smaller leaves and is less cold-resistant.

The fruit varies in shape from round to slightly oval and is somewhat smaller in size, averaging around 10 to 12 g. The rind is smooth and well coloured like Nagami but somewhat thinner and sweeter. It has more segments (six or seven) with slightly more seeds: typically three to six. Opinions differ on which variety is the better quality; often they are indistinguishable.

Meiwa or Large Round Kumquat

(Neiha, Jindan)

(Fortunella crassifolia)
[*F. margarita* × *F. japonica*]

The Meiwa kumquat is little known outside the Orient, but it is generally considered to be of much better eating quality than the other varieties, Nagami and Marumi. It is grown widely throughout southern China and in south-western Japan, where it is also known as Neiha kinkan.

The tree is similar to Nagami and Marumi, except that it is thornless and more cold-resistant than either.

The fruit is slightly oval to round and typically 30 × 25 mm, has a smooth but very thick rind – twice as thick as the other two varieties – and very little juice. There are usually seven segments, the walls of which are thickened, and it has few if any seeds.

Because of its lack of juice, thicker albedo and fewer seeds, Meiwa is much the sweetest flavoured of the kumquats. There are an estimated 2,000 tons produced annually in Japan.

LIMEQUATS

Eustis

Like its sister variety Lakeland, Eustis resulted from a cross between Mexican or Key lime and Marumi kumquat made by W. T. Swingle in Florida in 1909. The trees are usually of small size, and have an open growth habit.

The fruit is of similar size to its kumquat parent, is roundish-oval in shape, the rind is smooth and light yellow when fully mature and it has about eight very small seeds. It is moderately juicy considering its small size, and is very acid. It has a sweetly flavoured rind, free from the bitterness of the lime rind.

Eustis Limequat

Lakeland

Fruit size is noticeably larger than Eustis, although the shape, peel texture and acidity are the same. It is a deeper yellow when mature and has fewer seeds, averaging about five per fruit.

THE SOUR OR BITTER ORANGE

Citrus aurantium

From its region of origin in south-east China and northern Burma, the sour orange spread through Japan and India. With the expansion of the Arab Empire in the Near East, sour oranges were planted throughout the region, eventually along the east and north coasts of Africa, and in southern Spain during the tenth century and well ahead of the sweet orange in 1450.

Sour or bitter oranges are of course both sour *and* bitter, the former due to the highly acidic juice and the latter to the bitter compound neo-hesperidin. They are known by several names:

Sour, bitter or Seville orange	English
Naranja Amargo	Spanish
Arancio Amaro	Italian
Bigarade or Orange Amère	French
Daidai	Japanese
Taitai	Chinese

Seville orange is probably best known for its fruit, used especially for the manufacture of marmalade. However, although there are other uses for extracts from the fruit for flavouring soft drinks and liqueurs such as Cointreau and Curaçao, it is for its importance as a rootstock that sour orange is best known to citriculturists.

The tree has a more upright growth habit and is thornier than the sweet orange. Moreover, it is more cold-hardy and can withstand adverse environmental conditions far better. It makes a handsome specimen tree, producing abundant fragrant blossoms and deep reddish-orange fruit which are capable of hanging on the tree for many

SOUR ORANGE TREES ARE USED TO LINE THE STREETS, AS HERE IN REHOVOT, ISRAEL

months. These attributes have been appreciated for centuries in private gardens and in public places. Throughout many towns and cities along the Mediterranean coast and in parts of Arizona and California and elsewhere in the world sour orange trees are employed to line the streets to give shade and effective decoration.

Sour orange flowers are used in the manufacture of oil of neroli which is used in the perfume industry. In Marrakech, Morocco, where streets and avenues are lined with sour orange seedlings, manufacturers buy from the municipality the right to pick the flowers in March and April.

COMMON BITTER OR SOUR ORANGE

This is the common bitter orange widely used as a rootstock and grown principally in Spain for marmalade production.

SEVILLE

(Daidai, Taitai)

The principal variety grown in Spain is the Sevillano which is in fact a group of selections made for their vigorous growth, freedom from thorns and good productivity. Unlike most other citrus varieties (except West Indian lime), Seville orange trees are raised as seedlings and are not budded onto a rootstock.

Known as Seville Sour or Malaga Bitter, the fruit is medium to large in size, roundish in shape but slightly flattened and depressed at the ends, and with a thick, slightly rough (sometimes very rough) rind of uneven texture.

Well coloured when fully mature, the fruit is harvested in January and February in Seville and Malaga, Spain. The segment walls are tough and the fruit is very seedy. Too bitter and too acidic to be eaten fresh, the fruit is processed primarily for marmalade but also for its rind oil and juice. When the leaves are crushed, they have a distinctive and pleasant aroma. Oil of petitgrain is distilled from them, and oil of neroli from the flowers.

Malaga Bitter oranges usually arrive on the market a little ahead of the Seville Sour, but are grown from the same selections although some people maintain that fruit from the Seville region has a more distinct flavour due probably to the slightly different climatic conditions of the two localities.

At one time much of the production of Seville oranges was shipped fresh, particularly to Dundee in Scotland, for manufacturing into marmalade. However, in recent years Seville orange production has declined as marmalade consumption has fallen, and much of the preliminary processing is now done in Spain before being exported in a preserved form. The annual production of Seville oranges presently stands at around 14,000 tons.

The sour orange is produced on a limited scale in Japan with an estimated annual production of 2,000 tons and it is known as Daidai. It is used primarily for flavouring foods as well as for decorative purposes. In China, where it is called Taitai, the fruit is used in both the fresh and dried forms and the dried flowers are mixed with tea.

SEVILLE SOUR ORANGE

CHINOTTO

VARIANT BITTER ORANGES

BOUQUETIERS

While flowers of the Seville orange are used for the production of oil of neroli, the real neroli industry in France and North Africa makes use of special varieties of sour orange, called Bouquetiers, of which there are several types:

Bouquetier à grandes fleurs (Large flowered sour orange)

Bouquetier de Nice à fleurs doubles (Double flowered sour orange)

and to a lesser extent

Bigaradier de Grasse (Sour orange of Grasse)

Bouquetier à fruits mous (Soft fruited sour orange)

MYRTLE-LEAF ORANGE

CHINOTTO

(Citrus myrtifolia)

This sour orange type, also known as the Myrtle-leaf orange, makes a small but highly decorative ornamental shrub. It is slow growing, and the thornless branches have a weeping habit. The internodes are so short that the leaves are crowded and compact. The fruit is small, slightly flattened with a pebbled rind texture, and is set in great numbers, producing a most attractive small tree. The fruit remains on the tree for long periods.

The Chinotto was once an important fruit for processing into glacé fruit and was one of the components of the famous gift boxes of crystallised fruits made in Apt, near Avignon, France, together with apricots, cherries, etc. Once grown in the Savona area of Italy, it has now disappeared and been replaced by small seedless Clementines.

BERGAMOT

(Citrus bergamia)

The origin and genetic background of the bergamot is unknown but it has been well established in the Mediterranean for many centuries. It most probably has a sour orange as one parent and, it has been suggested, the Palestine sweet lime as the other.

Bergamot trees are moderately vigorous, spreading in habit, thornless and, when fully grown, are of medium-small size. They are grown on sour orange rootstock and bloom only once a year in April and May.

The fruit is almost round in shape, small to medium in size and often has a persistent style. Yellow in colour and with a smooth or very slightly pebbled rind, it might at first sight be mistaken for a lemon although for some inexplicable reason it has misleadingly been referred to as bergamot orange! On handling and scratching the rind, the oil released has a beautifully perfumed and unique aroma.

The fruit is particularly firm, the rind moderately thick, the flesh straw-coloured and of very acidic and bitter flavour. It has a high juice content and is usually very seedy.

BERGAMOT
Grown only for its distinctively perfumed rind oil

Bergamot production is restricted almost entirely to the coastal region in Calabria Province at the southernmost part of mainland Italy, although a minute industry also exists in Turkey and the Ivory Coast in West Africa. The Italian industry has been declining for many years, and today it covers less than 1,600 ha with average production of around 10 tons/ha.

The oil was formerly extracted by hand in the old and traditional manner of rubbing the fruit against a hard surface and collecting the expelled rind oil by means of a sponge. The whole process is extremely tedious, since the oil yield averages only 5 to 6 kg per ton of fruit. Nowadays the process has been mechanised. The fruit is rasped to release the oil under a water spray. The oil is then separated from the water centrifugally. Total annual world production is probably no more than 100 tons.

Bergamot oil is an important component of two very well-known products: perfume and tea.

The oil forms the basis of eau de cologne, first developed in Cologne, Germany, in 1676 by an Italian immigrant, Paolo Feminis, but not commercialised until 1709 by his son-in-law, Gian Maria Farina. Bergamot petitgrain oil is another but less important product distilled from the leaves and twigs. Tobacco is sometimes perfumed with bergamot oil.

The well-known semi-scented English-style tea, Earl Grey, is prepared by spraying the blended tea with oil of bergamot – the secret is said to have been presented to the second Earl Grey by a Chinese Mandarin around 1830.

There are three recognised varieties in Italy which differ mainly in their fruit yield and oil quality characteristics.

CASTAGNARO

This old variety has very vigorous tree growth but tends to alternate-bearing. The fruit is large and very rough, but the oil yield is below average and of poorer quality than the other two varieties. Castagnaro is usually harvested late, in January to February.

FANTASTICO

A recent variety, Fantastico accounts for all new plantings and 70 per cent of Italy's present bergamot production. The tree is fairly vigorous, yields well and has only a slight tendency to alternate-bearing. The fruit is of medium size, averaging about 130 g, with a rough rind texture and yielding a good quality oil. A midseason variety, it is harvested in December to January.

FEMMINELLO

Femminello is an old variety contributing about 20 per cent of the total crop. The tree lacks vigour, and yields poorly. The fruit is small, smooth, and has a high oil content of good quality. This variety matures earlier than Fantastico.

CITRUS ROOTSTOCKS

Because of the significant influence which rootstocks have on the quality and appearance of the fruit of the scion variety, as well as other important horticultural characteristics, no review of citrus varieties is complete without outlining some of the rootstocks' main features.

Of greatest importance are tree growth and longevity, fruit yield and quality. At one time producers were primarily concerned with maximising yield on vigorous, healthy trees, but increasingly they have become more aware that fruit size and quality also have an important part to play when choosing rootstocks as well as improved scion varieties or selections.

In selecting a rootstock, careful consideration must be given not only to yield and fruit size and quality but also the tree's ability to withstand drought, cold, salt and alkalinity. In addition, it is the rootstock's interaction with the scion part of the tree which enables the tree as a whole to withstand the adverse effects of pests and diseases.

Until about 1840 most citrus was produced from seedlings, due partly to the ease with which seeds could be transported as citriculture expanded worldwide.

It was not until root rot or foot rot (*Phytophthora*) was recognised as a major disease of trees in the Azores that the practice of budding trees onto tolerant rootstocks began. As phytophthora spread to all major orange-producing countries, so too did the need to adopt budded trees as a means of combating the disease.

Initially sour orange was used exclusively as the rootstock, but after trees in South Africa and Aus-

PROPAGATION BY BUDDING

Above left An incision is made into the seedling rootstock, and the scion bud inserted. The bud sticks are held in the budder's other hand

Above centre After inserting, the bud is wrapped

Above When the bud has united with the stock and the wrapper is removed, the stock is pruned back. Later growth starts from the inserted bud

Left The union between Shamouti scion and Volkamer lemon rootstock is not completely compatible

tralia declined with a disease later identified as citrus tristeza virus (CTV), rough lemon was adopted with considerable success.

As tristeza spread to California, Florida and some South American countries, and later to the Mediterranean region, the search began for rootstocks other than rough lemon which would tolerate the virus. As a result there is today a range of rootstocks available to producers. Some are natural citrus species such as sour orange and rough lemon mentioned earlier, as well as sweet orange, trifoliate orange, Cleopatra mandarin, Volkamer lemon, Palestine sweet lime, alemow (macrophylla) and Rangpur. In addition, there are several man-made hybrids which include citranges such as Troyer and Carrizo (sweet orange crossed with trifoliate orange) and citrumelos (sweet orange crossed with grapefruit) like Swingle citrumelo.

There follows a brief outline of the more important rootstocks and their characteristics as they affect tree health, yield, fruit quality and size and other horticultural characteristics.

SOUR ORANGE

(Citrus aurantium)

At one time the main rootstock used worldwide and still widely employed wherever tristeza is not a problem, sour orange imparts moderate vigour to the tree, giving good yields of high quality fruit. Not only is the fruit high in total soluble solids (sugar content) with moderate to high acid levels but it can be stored on the tree for longer periods than some other rootstocks without a marked deterioration in quality. Fruit size is usually good, as is the ascorbic acid (vitamin C) content.

Sour orange's ability to grow in wet soil conditions is due to some extent to its tolerance of phytophthora root rot. It is also more tolerant to salt and performs well on calcareous soils but does not yield well on sandy soils compared with some other rootstocks.

Together with trifoliate orange and Swingle citrumelo, sour orange imparts the best cold-hardiness of all rootstocks, and the incidence of 'blight' in Florida is low where trees are on sour orange. However, it is susceptible to both citrus and burrowing nematodes.

Trees on sour orange are unaffected by exocortis or xyloporosis, nor is its use for lemons precluded in tristeza-affected areas since the hypersensitive reaction of the lemon at the site of infection prevents the virus from becoming established. Any sprouts from below the budunion must of course be removed immediately to prevent direct infection of the sour orange rootstock. In China, a variant sour orange Gou-tou is said to be tolerant to tristeza but this has yet to be confirmed elsewhere.

Sour orange is used extensively in many countries. For example, in Mexico where tristeza is not found, it is employed almost exclusively except for Mexican limes. Much of the orange and mandarin crops in Italy, Greece, Turkey and North African countries are produced on sour orange-rooted trees. In Spain the use of sour orange as a rootstock for all varieties is prohibited by law, except for lemons. In Argentina lemons are grown on sour orange as well as Volkamer lemon rootstock. On the heavy clay soils of China along the south-east coast sour orange is the most preferred rootstock.

ROUGH LEMON

ROUGH LEMON

(Citrus jambhiri)

Rough lemon imparts good vigour and results in trees of large size, particularly on sandy soils in warm, humid regions. However, in addition to producing heavy yields of large size fruit, the quality is invariably adversely affected compared with other rootstocks. The fruit often has thicker, coarser rind, a lower juice content and the juice itself is low in sugars and has a low solids to acid ratio. There is often a marked tendency to produce granulated fruit, especially when the trees are young.

In addition to being tolerant to tristeza, trees on rough lemon are very drought-tolerant because of their extensive well-developed root system. Moreover, they are moderately salt- and alkaline-tolerant and are not greatly affected by the exocortis and xyloporosis viroids.

Although rough lemon is not tolerant to phytophthora foot rot and is susceptible to citrus and burrowing nematodes, the vigour of trees grown on it enables them to perform well in replant situations.

ALEMOW OR MACROPHYLLA

As well as producing inferior quality fruit, trees on rough lemon are less cold-hardy than on most other rootstocks, although cold-damaged trees are able to recover quickly due to their good vigour. It is intolerant to 'blight' in Florida, with the result it is little employed there on new plantings.

Rough lemon, however, is still an important rootstock on existing plantings in Florida and widely used in southern Africa, India, Colombia, Arizona and parts of Australia.

ALEMOW OR MACROPHYLLA

(Citrus macrophylla)

Alemow (more commonly referred to as macrophylla) is thought to be a hybrid of the citron and pummelo originating from the island of Cebu, Philippine Islands. Trees on this rootstock are exceptionally vigorous and have outstanding productivity from an early age. They are very tolerant of high soil levels of chloride, calcium and boron.

However, it has some serious limitations: sensitivity to cold and susceptibility to 'blight' and no tolerance to nematodes. It also produces trees which lean to one side.

Trees on Alemow are susceptible to tristeza and xyloporosis. With the exception of lemons, the internal quality of most varieties is poor on macrophylla rootstock.

In Spain and California, two of the world's principal lemon producers, macrophylla has become a very popular rootstock. Around 75 per cent of new Spanish lemon plantings are on macrophylla and it seems to be particularly well suited to Eureka since it increases yield and early bearing and produces larger fruit of finer rind texture than on sour orange. The Fino (Primofiori) variety yields are also increased but maturity is also advanced, which is a disadvantage; and increased fruit size and cold susceptibility of the Verna are serious adverse characteristics.

It is not recommended for Eureka lemons in Australia.

SWEET ORANGE

(Citrus sinensis)

At first it may seem strange that the sweet orange itself is useful as a rootstock for sweet orange varieties as well as for mandarins and lemons. However, sweet orange as a rootstock displays many good characteristics: scion varieties grow well and produce heavy crops of good sized, fine quality fruit.

They are moderately salt-tolerant and cold-hardy, and tolerant to exocortis, xyloporosis and tristeza virus, but have several shortcomings: they are unsuited to calcareous soils, susceptible to phytophthora foot rot on heavy soils, and their shallow root system makes them vulnerable to drought on sandy soils. In addition, most sweet orange rootstocks are susceptible to both citrus and burrowing nematodes. Where root rot can be controlled by the use of fungicides and when adequate irrigation is provided on light soils, sweet orange is worthy of evaluation as a potential rootstock.

Sweet orange was once a popular rootstock in California until tristeza became a problem, and is now no longer used. In Australia it is still important except on heavy soils on some inland irrigated areas. It is of minor importance in Argentina and in Brazil, except in Sao Paulo State where Rangpur is almost exclusively employed.

TRIFOLIATE ORANGE

(Poncirus trifoliata)

The trifoliate orange is an important rootstock in several parts of the world and so too are some of its progeny. When it is crossed with sweet orange, the citranges produced are equal to, or more important than, the trifoliate orange itself.

Its progeny when crossed with grapefruit (citrumelos) have also attracted attention in the past two decades as having better potential as rootstocks.

Trifoliate orange is regarded as being a semi-dwarfing rootstock, conferring very good cold-hardiness to trees budded on it and producing fruit of outstanding quality. It is susceptible to exocortis, while its great cold-hardiness is best expressed in relatively cold areas such as in parts of Japan.

TRIFOLIATE ORANGE

Trifoliate orange trees themselves are deciduous and can be extremely cold-hardy – one growing in my garden in southern England has experienced very hard freezes of −25°C unscathed. It is an attractive small shrub extensively grown in Japan and China.

In addition to the outstanding rootstock characteristics described above, trifoliate orange is also an excellent rootstock for replantings because it is resistant to phytophthora root rot and citrus nematode, as are some of its hybrids.

Trees on trifoliate orange are unaffected by tristeza and xyloporosis, but the former virus sometimes causes stem-pitting. It is not tolerant to the burrowing nematode and is susceptible to 'blight' in Florida and to a similar disease, marchitamiento repentino, in Argentina and Uruguay.

There are several strains of trifoliate orange, somewhat different with regard to morphology and their effect on tree size. One selection with a marked tendency to produce dwarf trees is named

Flying Dragon and is considered by some to be a probable hybrid. Its use in high-density plantings like those successfully practised with apples on dwarfing rootstocks is being evaluated with various citrus scions in some countries.

CITRANGES

Originating as hybrids of trifoliate orange and sweet orange, there are now several varieties of citranges. Most widely grown are Troyer and Carrizo; less well known ones are Morton, Rusk and Benton.

The initial stimulus for citrange production was the persistent frosts in Florida and the need to combine the cold-hardiness of the trifoliate orange with the good traits of sweet orange.

Troyer and Carrizo resulted from crosses of trifoliate orange with navels, and are very similar in many respects. Trees on both grow fairly vigorously on a range of soil types but have poor salt-tolerance and are sensitive to calcareous soils and are very susceptible to exocortis.

Fruit quality is similar to that of the sour orange, but any creasing of the rind is exacerbated when citrange rootstocks are used. Eureka lemons are incompatible with citranges except the Australian variety, Benton.

Trees on Troyer and Carrizo are resistant to xyloporosis but susceptible to 'blight' and are less cold-hardy than those on Cleopatra, trifoliate orange and sour orange – Carrizo particularly so.

Troyer, and to a lesser extent Carrizo, has become a very important rootstock in many countries; for example, in Spain 70 per cent of new plantings (except lemons) are on these citrange stocks. Troyer is twice as popular as Carrizo but the latter is gaining in importance because it gives better yields of similar quality to Troyer.

In California, Troyer is now the predominant rootstock. Gradually limes in Mexico are being budded to Troyer and Carrizo, while the use of Troyer is becoming more widespread in Israel, southern Africa and Australia. Carrizo is an important stock in Florida, while in California it is being used on an increasing scale.

TROYER CITRANGE

YUZU

(Citrus junos)

In the early years of the Japanese Unshiu (satsuma) industry, Yuzu was the predominant rootstock but was later replaced by the trifoliate orange which confers better cold-hardiness, a factor of considerable importance there, as well as producing smaller size fruit of better quality.

Today Yuzu is still used on a reduced scale for other varieties; it is the recommended rootstock in south-west Japan for Tankan, and is also used for inarching declining Unshiu trees throughout other areas.

Yuzu is cold-hardy and tolerant to phytophthora foot rot and tristeza. It is produced not only for its seed as a rootstock source but also for its fruit which is widely used in Japan as a garnish.

PALESTINE SWEET LIME

(Citrus limettioides)

A description of this variety appears on page 103. Quite possibly a hybrid of four species including the lemon and Mexican lime, Palestine sweet lime has been of significant importance as a rootstock only in India and the Middle East, particularly Israel, but there it is no longer propagated, although a few older Shamouti plantings on it still exist.

Trees on Palestine sweet lime perform best when the area is free from virus diseases, producing good sized fruit but of little better quality than those on rough lemon.

PALESTINE SWEET LIME

CLEOPATRA MANDARIN

(Citrus reshni)

Cleopatra mandarin has both good and poor traits as a rootstock, and has not assumed more than minor importance worldwide but increasing interest is being shown in some countries.

Although attaining good size, trees on Cleopatra are low-yielding, and late in starting to bear and reaching peak production, especially with oranges, much less so with mandarins. They also produce fruit of small size although of outstanding quality.

Tolerant to exocortis, xyloporosis and tristeza virus, cold-hardy and salt-tolerant, trees of Cleopatra grow well on calcareous, sandy and heavy clay soils, showing good phytophthora tolerance but much less dependably so than those of sour orange.

In Florida an estimated 10 per cent of new plantings are on Cleopatra, particularly the large-fruited mandarin varieties Temple, Nova and Honey tangerine (Murcott), and early and midseason oranges – Hamlin and Pineapple – but not Valencia.

CLEOPATRA MANDARIN

In Spain around 20 per cent of nursery trees – oranges, mandarins and grapefruit – are budded to this rootstock. It is of some importance in India and of considerable significance in north-west Argentina where it shares equal importance with Rangpur. An interest is shown in Cleopatra in Brazil for 'blight' tolerance and in Israel because of tristeza.

RANGPUR

(Citrus limonia)

Not related to the Mexican lime (*C. aurantifolia*) nor the Persian lime (*C. latifolia*), Rangpur (sometimes referred to as Rangpur lime) is probably a hybrid acid mandarin with either the rough lemon or sour orange involved in the parentage.

Trees on Rangpur have vigour similar to those on rough lemon and are just as productive, yielding fruit somewhat smaller in size but of only moderate quality which is little better than rough lemon. They start to bear early and excel in salt-tolerance. However, the trees have the same lack of cold-hardiness and are phytophthora-sensitive.

Although showing good drought tolerance due to a well-developed root system, the Rangpur is not tolerant to the citrus or the burrowing nematode, and it is at present suffering heavy tree loss due to 'blight' ('declinio') in Brazil. Rangpur-rooted trees are tolerant of salty and calcareous soil conditions as well as tristeza, but are susceptible to the exocortis and xyloporosis viroids.

Almost all of the enormous orange crop of Sao Paulo State, Brazil, is grown on Rangpur (Limão cravo), which makes it very vulnerable to any as yet unknown disease or new strain of an existing one.

Elsewhere other than in Argentina, Rangpur is now rarely employed.

RANGPUR

VOLKAMER LEMON

(Citrus volkameriana)

Believed to be a hybrid between the lemon and sour orange, and of Italian origin, the Volkamer lemon (commonly referred to as Volkameriana) has recently attracted attention as a potential rootstock because of its tolerance to mal secco disease and phytophthora root rot, as caused by *P. parasitica*.

Trees on Volkamer lemon are vigorous and very productive but fruit quality is little better than rough lemon, although it does confer better cold-hardiness. It is tolerant to tristeza virus and somewhat to salinity and, like sour orange, grows well on calcareous soils.

VOLKAMER LEMON

However, it is susceptible to citrus and burrowing nematodes and to 'blight'. Volkamer lemon is recommended for lemons in Italy, particularly for the Monachello variety since it presents none of the incompatibility problems of sour orange.

It has been employed on a very limited scale in Argentina, Brazil, Israel, Florida and South Africa.

SWINGLE CITRUMELO

A hybrid between trifoliate orange and grapefruit, named after Walter T. Swingle, the famous citriculturist, this rootstock has received increasing attention in the past two decades. It produces fruit of outstanding quality equalling that produced on sour orange while conferring just as good cold-hardiness on the tree.

Yields of many varieties are better than those on sour orange. The trees were thought to be tolerant to exocortis and xyloporosis but this is now in doubt. They are tristeza-tolerant as well as being resistant to phytophthora root rot and the citrus nematode, but they are severely chlorotic on calcareous soils.

Intensive evaluation of Swingle citrumelo is being undertaken in Florida on a range of varieties, while in California it appears to hold most potential as a rootstock of Lisbon lemons, but in Spain it is incompatible with the major lemon varieties and makes a poor budunion with Clementines, especially Oroval. For these reasons it is no longer regarded as having much potential in Spain. Its main fault is overgrowing by the scion, often leading to early decline with oranges and mandarins. It is also being tested in Israel.

CITRUS TAIWANICA

(Citrus taiwanica)

Known in Japan as Nanshôdaidai, *C. taiwanica* was at one time of some promise as it was reported to be tristeza-tolerant, but fruit quality of most varieties was poor on this stock and it is of little interest today.

CITRUS TAIWANICA

Further Reading

Advances in Fruit Breeding J. Janick and J. N. Moore (Purdue University Press) 1975
Citrus Fruits H. H. Hume (Macmillan Co.) 1966
Citrus Growing in Australia F. T. Bowman (Angus & Robertson) 1956
Citrus Growing in California K. W. Opitz and R. G. Platt (University of California) 1969
Citrus Growing in Florida L. W. Ziegler and H. S. Wolfe (University Presses of Florida) 1975
Citrus Science and Technology (Vol 2) S. Nagy, P. E. Shaw and M. K. Veldhuis (Avi Publishing Co Inc) 1977
Cultivation of Neglected Tropical Fruits with Promise (Part 3: The Pummelo) F. W. Martin and W. C. Cooper (Agric. Res. Service, USDA) 1977
Florida Citrus Varieties D. P. H. Tucker and C. J. Hearn (University of Florida) 1982
Fresh Citrus Fruits W. F. Wardowski, S. Nagy and W. Grierson (Avi Publishing Co Inc) 1986
In Search of the Golden Apple W. C. Cooper (Vantage Press) 1982
Les Agrumes J. C. Praloran (G-P Maisonneuve & Larose) 1971
The Citrus Industry (Vol 1) W. Reuter, H. J. Webber and L. D. Batchelor (University of California) 1976
Trattato di Agrumicultura P. Spina (Edagricole) 1985
Rootstocks for Fruit Crops R. C. Rom and R. F. Carlson (John Wiley & Sons) 1987
Variedades de Agrios Cultivades R. Bono, J. Soler and L. Fernandez de Cordova (IVIA Moncada, Valencia) 1985

INDEX